Recent Advances in
Catalytic Materials

MATERIALS RESEARCH SOCIETY
SYMPOSIUM PROCEEDINGS VOLUME 497

Recent Advances in Catalytic Materials

Symposium held December 2–4, 1997, Boston, Massachusetts, U.S.A.

EDITORS:

Nelly M. Rodriguez
Northeastern University
Boston, Massachusetts, U.S.A.

Stuart L. Soled
Exxon Research and Engineering Company
Annandale, New Jersey, U.S.A.

Jan Hrbek
Brookhaven National Laboratory
Upton, New York, U.S.A.

Materials Research Society
Warrendale, Pennsylvania

CAMBRIDGE UNIVERSITY PRESS
Cambridge, New York, Melbourne, Madrid, Cape Town,
Singapore, São Paulo, Delhi, Mexico City

Cambridge University Press
32 Avenue of the Americas, New York NY 10013-2473, USA

Published in the United States of America by Cambridge University Press, New York

www.cambridge.org
Information on this title: www.cambridge.org/9781107413528

Materials Research Society
506 Keystone Drive, Warrendale, PA 15086
http://www.mrs.org

© Materials Research Society 1998

First published 1998
First paperback edition 2013

Single article reprints from this publication are available through
University Microfilms Inc., 300 North Zeeb Road, Ann Arbor, MI 48106

CODEN: MRSPDH

ISBN 9781107413528 Paperback

TABLE OF CONTENTS

*Invited Paper

*Invited Paper

*Invited Paper

*Invited Paper

PREFACE

Heterogeneous catalysis is undergoing a dramatic change that is being driven by an increased awareness of environmental issues, the development of new high-tech materials, and the need to develop not only more energy-efficient processes, but also catalysts that exhibit improved performance over that observed with current systems for the production of specialty chemicals. The traditional concept of a catalyst system consisting of finely divided metal particles dispersed on an inert ceramic support has encountered increased challenges. Researchers are now taking a much closer look at not only the chemical and physical phenomena that occur at the metal particle/support interface, but also the detailed structural and crystallographic features of both components. A further aspect of interest in this emerging approach to the study of catalytic materials is the notion that the electronic properties of the substrate can be of extreme importance since they can influence not only the activity, but also the selectivity, of supported metal particles.

The Materials Research Society has established a series of symposia dealing with catalytic materials in which scientists from very different disciplines have made contributions that brought a fresh insight to the field. A sample of those new ideas are condensed in this volume. The objective of our symposium was to provide a forum for an exchange of diverse knowledge and experience of researchers in various areas of materials science which could have an impact in heterogeneous catalysis.

New and exciting approaches were presented at the symposium on 'Recent Advances in Catalytic Materials.' A number of new techniques have enabled researchers to obtain information on the physical and electronic properties of metal clusters supported on metal oxides. Technologies that have been in existence for the processing of electronic materials are now being incorporated into the area of catalysis to produce novel nanostructured catalytic systems with extraordinary properties. Although support materials have been traditionally insulators, it is now being recognized that the use of structurally tailored and electrically conductive graphite nanofiber and fullerene-based supports can induce some unexpected changes on the catalytic behavior of small metal particles. The concept of activating catalyst surfaces by manipulation of the electronic properties of the substrates was perhaps one of the most fascinating concepts presented at this symposium. Surface acoustic waves and resonance oscillations have been found to have a significant effect on the arrangement and electronic structure of surface atoms, and therefore can alter not only the activity, but also the selectivity, of a catalytic system. Since the discovery of the unusual catalytic properties of zeolites, oxide catalysts are now being carefully crafted so as to produce materials with extremely well-defined structural properties, where the manipulation of pore size is crucial to the activity and selectivity of catalysts for a variety of chemical reactions.

Nelly M. Rodriguez
Stuart L. Soled
Jan Hrbek

August 1998

MATERIALS RESEARCH SOCIETY SYMPOSIUM PROCEEDINGS

MATERIALS RESEARCH SOCIETY SYMPOSIUM PROCEEDINGS

Prior Materials Research Society Symposium Proceedings available by contacting Materials Research Society

Part I

Oxide Catalysts

TEXTURAL AND ACIDIC PROPERTIES OF MIXED ALUMINA-SILICA OXIDES PREPARED WITH COMMERCIALLY AVAILABLE SOLS

STEVEN J. MONACO AND EDMOND I. KO
Department of Chemical Engineering, Carnegie Mellon University, Pittsburgh, PA 15213-3890.

ABSTRACT

In this study we have used commercially available preformed sols as building blocks to systematically explore the effects of composition, particle size, and packing on the textural and acidic properties of alumina-silica. We have prepared single oxides and alumina-silica mixed oxides with varying Al:Si atomic ratios using commercial sols from Vista Chemical Co. (alumina) and Eka Chemicals, Inc. (silica). Simple particle packing models based on the structure and experimentally determined particle size distributions of the sols explain the textural and acidic properties of both the single and mixed oxides. Comparisons with aerogels prepared from alkoxides show that materials with different atomic-scale homogeneity can be obtained. This continuum of precursor sizes from monomer through colloid allows a measure of control over textural and acidic properties in the mixed oxides, even at a fixed composition. These results show that systematic studies using preformed sols add insight into the effect of preparation upon catalytic materials.

INTRODUCTION

The distribution, or extent of mixing, of the individual components in a mixed oxide can be manipulated using sol-gel synthesis. Such a synthesis allows materials with different homogeneity to be obtained at a single composition. Control of the "goodness of mixing" of a sample allows some control over textural and acidic properties of catalytic interest such as thermal and hydrothermal stability or pore structure. One approach to controlling homogeneity is to use precursors of different sizes, and in this study we have used commercially available preformed sols of varying sizes as our building blocks. Preformed sols are less well studied and more difficult to manipulate than the commonly used alkoxide sol-gel precursors, but offer the advantages of lower cost and ease of handling. Preformed sols also allow the effect of sol particle size to be studied more easily than is possible in an alkoxide-based synthesis.

We have chosen to work with the alumina-silica system, historically the most studied mixed oxide due to its many industrial applications. Mullite ($3Al_2O_3 \cdot 2SiO_2$) is the subject of many studies because it is the most stable compound within the alumina-silica binary system. Mullite is considered to be a promising candidate for use as a high-temperature structural material due to its chemical and thermal stability, high-temperature strength and creep resistance, low thermal expansion, low density, and low thermal conductivity.[1,2] Other potential applications for mullite include its use as a substrate in multi-layer packaging for microelectronics and as infrared-transparent windows for high-temperature optical applications.[3] Aluminosilicates such as zeolites and clays also have useful catalytic properties such as thermal and hydrothermal stability, high activity, and suitable pore structures which make them useful for cracking catalysts and catalyst matrices.[4] However, other porous aluminosilicates with potential use in applications such as high temperature catalysis or separations have received comparatively little attention.[5]

The ceramics and glass literature contains many examples of studies of aluminosilicates formed from preformed sol systems. Such systems are often referred to as being "diphasic," where the gel consists of separate, possibly discrete silica and alumina rich "phases," each of them non-crystalline.[6] Diphasic gels are often prepared either using one sol and one solution (e.g., Al(NO₃)₃ + silica sol) or using two sols.[7] Studies of mullite formation and densification using various alumina and silica precursors to form diphasic gels abound in the ceramics literature.[8-17] However, none of these studies examined the effect of precursor sol size in any systematic or comprehensive fashion, nor were properties of interest to the catalytic community (i.e., textural properties, acidity, etc.) examined, since the intended result (maximum densification) was opposite to that desired for a catalytic application where porous materials are required. The few studies of diphasic, porous aluminosilicates examined from a catalytic perspective which exist include Sheng et al., who prepared mixed alumina-silica membranes by mixing sols prepared separately from aluminum tri-*sec*-butoxide (ATSB) and tetraethyl-orthosilicate (TEOS)[18] and a study of microporous aluminosilicates from aqueous silica sols performed by Chu and Anderson.[4] However, in neither of these studies were the properties of the precursor sols examined in detail.

In this study we have examined the properties of mixed oxide gels prepared from commercially available aqueous silica and alumina precursors. We have previously examined single oxide gels prepared from these same precursors.[19] The focus of our analysis of the alumina-silica samples has been to define their pore characteristics and acidic properties, since we are interested in these materials for catalytic applications. Our primary variables were precursor sol size (controlled by changing sol precursors) and the gel Al:Si atomic ratio (controlled by changing the relative amount of the alumina and silica sols used during preparation). One focus of this paper will be on a set of alumina-silica powders prepared from the same alumina and silica sols at 9:1, 1:1, and 1:9 Al:Si atomic ratios. It is instructive to examine the effect of changing the Al:Si atomic ratio because the many different potential applications for alumina-silicas may require a range of atomic ratios in order to obtain the desired properties. The second focus will be a comparison between three different samples prepared from sols and two alkoxide-derived aerogels, all prepared with a 1 Al:1 Si atomic ratio. This comparison will clearly show the effects of homogeneity on textural and catalytic properties in the alumina-silica system.

EXPERIMENT

Sample Preparation

Two colloidal aluminas provided by Vista Chemical Company were used: Dispal 11N7-12 and 23N4-20. Dispal 23N4-20 is an acid stable colloidal alumina, while 11N7-12 is stabilized at neutral pH. Three colloidal silicas, provided by Eka Nobel, Inc., were also used: Nyacol grades 215, 830, and 9950. Tables I and II respectively list the manufacturer's specifications for the alumina and silica sols.

Table III lists the samples prepared from commercial sols which are the focus of this paper, along with the main parameters used in their preparation. Details of the preparation of the alkoxide-derived aerogels referred to later can be found in Miller, Tabone & Ko.[20] The general preparation technique for the 1 Al: 1 Si sol-derived samples was as follows: The appropriate amounts of the silica and alumina sols (total volume = 15 mL) were pipetted into separate beakers with stir bars. In a separate beaker, a nitric acid plus deionized water solution was made such that the mixture of the sols and the acid solution would give the final concentrations listed

Table I. Colloidal aluminas available from Vista Chemical Co.

Dispal Grade	11N7-12	23N4-20
Nominal particle diameter, nm	415	90
Measured particle diameter, nm	191.1 (17.27 vol %) 849.3 (82.72 vol %)	13.3 (77.06 vol %) 52.5 (21.61 vol %) 372.0 (1.33 vol %)
Calculated particle diameter (383 K), nm	29.0	8.1
Al_2O_3, wt%	12	20
NO_3^-, wt%	0.02	0.38
Crystallite size (020), nm	20.4	5.6
Crystallite size (120), nm	38.4	10.3
pH	6.5-7.2	4.3-4.7

Table II. Colloidal silicas available from Eka Nobel, Inc.

Nyacol Grade	215	830	9950
Nominal particle size, nm	4	8-10	100
Measured particle diameter, nm	18.9 (86.4 vol %) 50.0 (13.6 vol %)	14.4 (93.51 vol %) 45.0 (6.49 vol %)	48.6 nm (100 vol %)
Calc. particle diam. (383 K), nm	8.6	13.3	40.2
SiO_2, wt%	15	30	50
Na_2O, wt%	0.83	0.55	0.15
pH	11	10.5	9

Table III. Alumina-silica gels: nomenclature and preparative parameters

Sample Name	Dispal Alumina Precursor	Nyacol Silica Precursor	Al_2O_3 Conc. (M)	SiO_2 Conc. (M)	HNO_3 Conc. (M)
90 Al_2O_3-8 SiO_2-(9 Al:1 Si)	23N4-20	830	1.00	0.22	0.14
90 Al_2O_3-8 SiO_2-(1 Al:1 Si)	23N4-20	830	0.84	1.68	0.05
90 Al_2O_3-8 SiO_2-(1 Al:9 Si)	23N4-20	830	0.19	3.50	4.52
90 Al_2O_3-100 SiO_2-(1 Al:1 Si)	23N4-20	9950	1.66	3.34	0.01
415 Al_2O_3-4 SiO_2-(1 Al:1 Si)	11N7-12	215	0.60	1.21	0.00

in Table III. The acid and water mixture was then added to the silica sol, immediately followed by the alumina sol. Gelation of this final mixture occurred almost immediately. Stirring was maintained during this procedure in order to maximize mixing between the sols. Stirring was stopped either when the vortex formed by the stir bar disappeared or 10 minutes had elapsed, whichever came first. The preparation of the 90 Al_2O_3-8 SiO_2-(9 Al:1 Si) and 90 Al_2O_3-8 SiO_2-(1 Al:9 Si) samples paralleled the preparation method used previously to prepare the major component as a single oxide.[19] For these mixed oxides, the alumina and silica sols were mixed together first, and then the appropriate amount of a nitric acid plus deionized water solution was added to give the concentrations listed in Table III.

The gels were dried in an oven at 383 K after a three hour aging period. The first hour of drying was at atmospheric pressure, and the final two hours were under vacuum at a pressure of 3.4 kPa absolute. The gels were then cooled and covered overnight. In the next heat treatment step, the gels were dried at 523 K under vacuum for an additional three hours, cooled, and ground to a powder which would pass through a 100 mesh sieve. The ground powders were calcined at 773 K for two hours in flowing oxygen (24 L/h). Subsequent calcinations at 1173 K, 1373 K, and 1473 K were performed on portions of the samples which had been previously heat treated to the prior temperature.

Characterization

Textural properties (BET surface areas, pore volumes, and pore size distributions) of the powder samples were determined by nitrogen adsorption performed on a Quantachrome Corp. Autosorb-1 gas sorption system. Before analysis, powders were outgassed at 473 K (383 K for powders previously heat treated to only 383 K) for three hours under vacuum.

Particle size distributions (see Tables I and II) of the silica and alumina precursor sols were performed by Particle Sizing Systems, Inc. using a NICOMP 370 submicron particle sizer. The NICOMP 370 measures particle size distributions using dynamic light scattering (DLS).

Diffuse reflectance infrared Fourier transform (DRIFT) experiments were performed using a Mattson Galaxy 5020 FTIR Spectrometer with a DTGS (deuterium triglycine sulfate) detector and a Harrick Scientific Praying Mantis Diffuse Reflection Attachment (DRA-2-MA1). Each IR spectrum was obtained by averaging 128 scans taken over a range of 400-4000 cm^{-1} with a 2 cm^{-1} resolution. Data were normalized with respect to the maximum peak within the broad structural vibration region (400-1000 cm^{-1}). Ex situ DRIFT spectra were taken using 2 wt% sample diluted in KBr. In situ pyridine adsorption experiments were performed using the Harrick Praying Mantis and Reaction Chamber apparatus. These DRIFT spectra were taken using 5 wt% sample diluted in KBr in order to enhance the pyridine signal. The samples were pretreated to 588 K under 3.0 L/h He flow at a heating rate of ~15-20 K/min, cooled to 373 K, and then exposed for 15 minutes to pyridine introduced via a bubbler. After the pyridine exposure, the samples were then heated to 713 K in a stepwise manner (373 K, 423 K, 473 K, 553 K, 623 K, 673 K, 713 K) with DRIFT spectra taken after five minutes at each step. These data were used to estimate relative populations of Lewis and Brønsted acid sites using the method of Basila and Kantner.[21]

In order to determine the dependence of the crystallization of silica on temperature as well as the transition temperatures of the alumina crystalline phases, X-ray diffraction (XRD) using a Rigaku D/Max diffractometer with Cu K_{α} radiation was performed on selected powders.

1-Butene isomerization experiments were carried out in a downflow, fixed bed reactor[22] on the samples which had been calcined to 773 K. Approximately 200 mg of calcined powder was loaded into the reactor. Samples were first pretreated at 573 K for 1 h in 50 sccm helium. The bed temperature was then lowered to 423 K and the reactor feed switched to a 100 sccm mixture of 5 mol% 1-butene in helium. The *cis*- and *trans*-2-butene reaction products were separated and quantified by gas chromatography. The steady-state activities reported later are based on these chromatography results after an on-stream time of 95 min.

RESULTS

Effect of Varying Composition

The three samples prepared from the Dispal 23N4-20 alumina and Nyacol 830 silica provide a basis to examine the effect of varying the Al:Si atomic ratio on the textural and acidic properties of alumina-silicas prepared from commercial sols. Figure 1 shows the surface area vs. heat treatment temperature and pore volume vs. heat treatment temperature data for the 9 Al:1 Si sample. Also shown are the data from a comparable alumina single oxide described elsewhere.[19] As can be seen from the figure, the effect of the silica on the textural properties is small over the entire temperature range. After 773 K heat treatment, the peak diameter of the pore size distribution was 6.9 nm, very close to the 6.3 nm of the pure alumina. At elevated temperatures (≥ 1173 K), the silica does help maintain surface area when compared to the pure alumina.

6

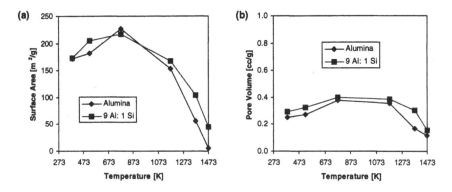

Figure 1. Surface area (a) and pore volume (b) versus heat treatment temperature for 9 Al: 1 Si powder prepared from Dispal 23N4-20 alumina and Nyacol 830 silica.

Furthermore, an examination of the DRIFT and XRD data show that the alumina transformation sequence (boehmite → gamma →delta → theta → alpha) is delayed when compared to the alumina single oxide prepared from the same Dispal 23N4-20 sol. For example, the γ- to δ-Al_2O_3 transformation is delayed from 1173 K to above 1373 K. The pyridine adsorption experiments showed that this sample had very few Brønsted acid sites, while those that were present were deactivated rapidly in the 1-butene isomerization reaction. No steady-state activity was observed.

Figure 2 shows the textural property data vs. heat treatment temperature for the 1 Al:1 Si sample prepared from the Dispal 23N4-20 alumina and Nyacol 830 silica. Also shown are data for the corresponding alumina and silica single oxides prepared from these sols.[19] This sample shows a number of interesting properties. There is a synergistic effect on surface area at this composition - the mixed oxide has higher surface area than either of the two single oxides. However, the pore volume is intermediate to the two single oxides at treatment temperatures less than 773 K, is higher than either at 1173 K, but then crosses below the alumina at 1373 K. The

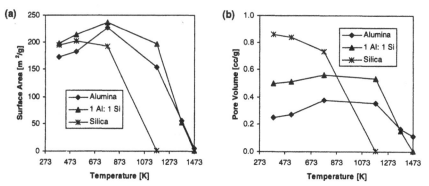

Figure 2. Surface area (a) and pore volume (b) versus heat treatment temperature for 1 Al: 1 Si powder prepared from Dispal 23N4-20 alumina and Nyacol 830 silica.

peak of the 773 K pore size distribution is intermediate to the two single oxides, with a peak pore diameter of 10.2 nm compared to 6.3 nm for the alumina and 15.7 nm for the silica. Again in the XRD data we see delays in the crystallization behavior of the component oxides - the silica transformation to cristobalite is delayed from 1173 K to 1473 K. The alumina also retains its gamma character to at least 1373 K, where the pure alumina is primarily δ-Al_2O_3. At 1473 K the intensity of the main cristobalite XRD peak at $2\theta = 22°$ dominates over the alumina peaks, making a definitive identification difficult. No mullite was detected by either the XRD or DRIFT techniques employed. A finite population of Brønsted acid sites was detected via pyridine adsorption, while the steady-state activity in 1-butene isomerization was small but measurable - 0.007 mmol/m^2/h on a surface area basis or 1.6 mmol/g/h on a mass basis.

The final sample in this series is a 1 Al:9 Si sample prepared from the same Dispal 23N4-20 and Nyacol 830 sols used for the other two samples. The textural property data vs. heat treatment temperature is shown in Figure 3. Also shown are data for the corresponding silica single oxide prepared from the Nyacol 830 sol.[19] At temperatures less than 773 K, the surface area and pore volume of the mixed oxide are lower than that of the pure silica. However, the mixed oxide retains its surface area and pore volume at higher temperatures much better than the pure silica. At 773 K, where the two curves intersect, the mixed oxide has a pore size distribution peak diameter (15.9 nm) very similar to that of the silica (15.7 nm). The source for this retention of surface properties is found in the DRIFT and XRD data - the pure silica crystallizes to cristobalite at 1173 K, while the mixed oxide remains amorphous up to 1473 K. One interesting observation is that the alumina was not detected by either the DRIFT or XRD techniques, indicating that the alumina "domain" sizes were very small or the signal from the alumina was dominated by the silica. This is curious because in the 9 Al:1 Si sample, the presence of both oxides was easily discerned. Much like the 9 Al:1 Si sample, the pyridine adsorption experiments showed that this sample had very few Brønsted acid sites (Lewis/Brønsted acid site ratio = 2.6), while no steady-state activity was observed in the 1-butene isomerization reaction.

Now that we have examined the properties of these three mixed oxides with varying Al:Si atomic ratio, it is necessary to illustrate how these results can be explained by invoking a relatively simple model. We have shown previously[19] that silica as a single oxide gel can be modeled as a loosely packed three dimensional array of non-porous spheres. Our preparation methods help to preserve the open structure of the gel through drying by forming strong

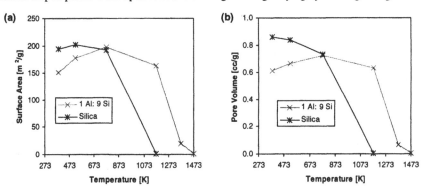

Figure 3. Surface area (a) and pore volume (b) versus heat treatment temperature for 1 Al: 9 Si powder prepared from Dispal 23N4-20 alumina and Nyacol 830 silica.

interparticle bonds in the wet gel. We have found good agreement between our *measured* particle size distributions and the particle sizes calculated from experimental textural property data using a spherical particle packing model (see Table II). The alumina must be treated somewhat differently. The alumina sols can best be described as aggregates of plate- or rod- like crystallites of boehmite. These crystallites are the "fundamental particle" which leads to the textural properties of a gel formed from these sols. As shown in Table I, the calculated particle sizes based on the measured surface area and pore volume are much closer to the crystallite sizes than to the particle size distributions. The structure of the aggregate itself seems to have little or no affect on the properties of the dried gel - the sols themselves show evidence that the aggregates tend to break up over time, and formation of a gel only hastens this process. With this knowledge of the behavior of the sols when forming single oxide gels, we can now explain the properties of the mixed oxides. Figure 4 shows pictorially how these mixed oxides might be described. The 9 Al:1 Si sample looks fundamentally like the pure alumina, and thus alumina's properties dominate. Since the silica particles remain intact and are relatively few in number, they are isolated from each other and mainly interact with the alumina crystallites. However, their presence is enough to interrupt the sintering and crystallization mechanism of the alumina, which leads to the delays in the alumina transformation sequence. But because they are few in number, little Brønsted acidity can be generated at the heterometallic Al-O-Si linkages between the two types of particles. The properties of the 1 Al:9 Si sample are also dominated by the major component, in this case silica. Because the alumina sol particles break up into the constituent crystallites, they are even more dispersed than the silica was in the 9 Al:1 Si powder. This could be why the alumina was not detected by XRD or DRIFT. These small crystallites reside in the pore volume between the silica spheres, leading to lower surface area and pore volume at heat treatment temperatures less than 773 K. The presence of the second oxide again interrupts the sintering and crystallization mechanisms of the major oxide, which leads to the retention of textural properties at the higher temperatures. These small alumina crystallites also are too few in number to create much Brønsted acidity. The 1 Al:1 Si sample has a structure

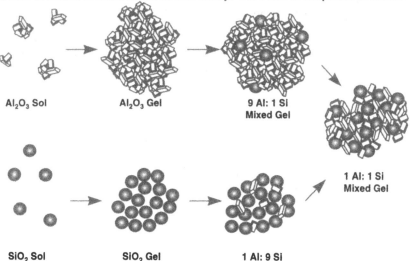

Figure 4. Model of Dispal 23N4-20 + Nyacol 830 alumina-silica mixed oxides.

intermediate to the 9 Al:1 Si and 1 Al:9 Si samples - there is enough of each oxide so that neither one's properties can dominate. Because the alumina crystallites and silica particles remain intact and segregated, the two crystallize separately and thus do not form mullite. However, each metal oxide delays the transformation of the other, thus we see the retention of textural properties at 1273 K. This effect disappears at higher heat treatment temperatures since there are sufficient amounts of each oxide that the inevitable loss of textural properties which accompanies their crystallization can only be delayed, not eliminated. One other effect of having relatively equal amounts of each oxide is that there are sufficient Al-O-Si linkages between the two to generate measurable Brønsted acidity and activity in the 1-butene isomerization reaction.

Effect of Homogeneity - Sol Particle Size

We have seen that the 1 Al:1 Si powder prepared from the Dispal 23N4-20 alumina and Nyacol 830 silica has properties that are in many ways intermediate to that of the single oxides prepared from the two sols separately. This sample could be described as mixing a "small" alumina plus a "small" silica. In order to explore the effects of homogeneity, we have chosen to fix our composition at a 1 Al:1 Si atomic ratio, and vary the size of the precursor sols. Two additional samples were prepared at this composition: the first using Dispal 23N4-20 alumina and Nyacol 9950 silica ("small" alumina + "large" silica), and the second using Dispal 11N7-12 alumina and Nyacol 215 silica ("large" alumina + "small" silica). This set of three samples allows us to examine the effects of homogeneity on alumina-silica mixed oxides prepared from commercial sols.

Textural property data for the Dispal 23N4-20 + Nyacol 9950 sample are shown in Figure 5, along with data for single oxide samples prepared from the two sols separately. The surface area is intermediate to that of the two single oxides, except at the highest heat treatment temperatures studied. The pore volume is very similar to that of the alumina single oxide over the entire temperature range. Much like the Dispal 23N4-20 + Nyacol 830 sample, the pore size distribution peak diameter at 773 K is intermediate to the two single oxides - 10.2 nm compared to 35.3 nm for the silica and 6.3 nm for the alumina. DRIFT and XRD both show delays in the crystallization of the silica and alumina. The silica remains amorphous and the alumina does not transform past γ-Al$_2$O$_3$, even at 1373 K. The steady-state activity in 1-butene isomerization for this sample was less than that of the Dispal 23N4-20 + Nyacol 830 sample discussed earlier -

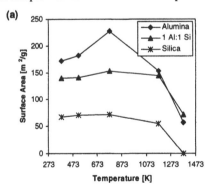
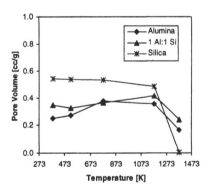

Figure 5. Surface area (a) and pore volume (b) versus heat treatment temperature for 1 Al: 1 Si powder prepared from Dispal 23N4-20 alumina and Nyacol 9950 silica.

0.002 mmol/m²/h on a surface area basis or 0.3 mmol/g/h on a mass basis. The Lewis/Brønsted acid site ratio measured by pyridine adsorption was 6.8.

Figure 6 shows the textural property data for the sample prepared from the Dispal 11N7-12 alumina and Nyacol 215 silica, along with data for single oxide samples prepared from the two sols separately. Yet again, we see that the textural properties are intermediate to that of the two single oxides at the lower heat treatment temperatures studied. However, at the higher temperatures (≥ 1173 K), the sample behaves more like the silica single oxide, albeit with delays in the textural property losses. The pore size distribution peak diameter at 773 K is intermediate to the two single oxides - 15.9 nm compared to 13.4 nm for the silica and 19.5 nm for the alumina. DRIFT and XRD both show delays in the crystallization of the silica and alumina. The silica remains amorphous and the alumina is a mixture of γ- and δ-Al_2O_3 at 1373 K. Only Lewis acid sites were detected in the pyridine adsorption experiment.

The effects of the relative particle/crystallite size on the properties of these 1 Al: 1 Si atomic ratio samples is evident in the experimental data we have presented. An additional factor we must also consider is the difference in density between silica and alumina – $\rho_{silica} = 2.2$ g/cm³ and $\rho_{alumina} = 3.0 - 4.0$ g/cm³, depending on its crystalline form. Figure 7 is a model which we can use to explain how these factors impact the structure of these mixed sol samples. We have already examined the Dispal 23N4-20 + Nyacol 830 sample ("small" alumina + "small" silica). In this case, the silica particles and alumina crystallites are similar in size. A rough calculation assuming spherical particles and using $\rho_{\gamma\text{-alumina}} = 3.2$ g/cm³ shows that there are approximately 3 crystallites of alumina per silica particle. This results in the enhanced stability of the textural

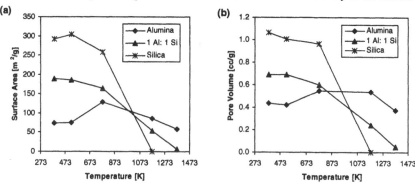

Figure 6. Surface area (a) and pore volume (b) versus heat treatment temperature for 1 Al: 1 Si powder prepared from Dispal 11N7-12 alumina and Nyacol 215 silica.

Dispal 11N7-12 + Nyacol 215　　　**Dispal 23N4-20 + Nyacol 830**　　　**Dispal 23N4-20 + Nyacol 9950**

Figure 7. Models of 1 Al:1 Si alumina-silica mixed oxides prepared from preformed sols.

properties and allows maximum interaction of the two metal oxides. This interaction is manifested in the catalytic activity and Brønsted acidity of this sample. In the Dispal 23N4-20 + Nyacol 9950 sample ("small" alumina + "large" silica), the silica particles are much larger than the alumina crystallites. In this case, the same calculation indicates that there are approximately 70 crystallites of alumina per silica particle, and thus this sample's properties are dominated by the alumina, even though the atomic ratio has been held constant. A similar effect occurs in the Dispal 11N7-12 + Nyacol 215 sample ("large" alumina + "small" silica), where the alumina crystallites are much larger than the silica particles. In this case there are approximately 65 silica particles per alumina crystallite, and thus the alumina crystallites are surrounded by a "sea" of the smaller silica particles. It is this "sea" of silica which dominates the sample's properties at the higher heat treatment temperatures. These examples show that we can control homogeneity on the nanometer scale of the precursor sol particles, and that homogeneity can have a dramatic effect on the properties of mixed oxides prepared from preformed sols.

Effect of Homogeneity - Sol vs. Alkoxide Preparation

In this final section of our experimental results, we will briefly compare the set of 1 Al:1 Si atomic ratio samples prepared from commercial sols to two alkoxide-derived aerogels prepared at this same composition. Details of the preparation of these aerogels can be found elsewhere.[20] One aerogel was prepared from prehydrolyzed tetraethylorthosilicate (TEOS) and aluminum *sec*-butoxide, while the second was prepared from a double alkoxide having the nominal molecular formula $(s\text{-OBu})_2\text{-AlOSi-(OEt)}_3$. As discussed previously,[20] this double alkoxide has "built-in" Al-O-Si linkages, which lead to aerogels that are exceptionally well-mixed on an molecular scale.

Figure 8 shows the textural property data for these five samples as a function of heat treatment temperature. The prehydrolyzed TEOS sample has properties which are very similar to that of the samples prepared from sols, while the double alkoxide sample initially has a much higher surface area and pore volume. The drop in textural properties for both of the alkoxide-derived samples is accompanied by the formation of mullite. This transformation is more complete in the case of the double alkoxide. In addition, some cristobalite is present in the prehydrolyzed TEOS sample (but not in the double alkoxide) at 1273 K and above. This is an

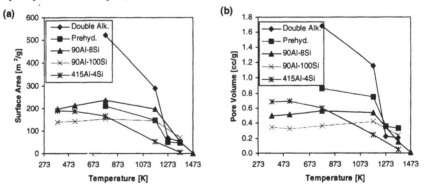

Figure 8. Surface area (a) and pore volume (b) versus heat treatment temperature for 1 Al: 1 Si powders prepared from alkoxides and preformed sols.

indication of degree of mixing present in the two alkoxide-derived samples. In the double alkoxide, the "built-in" Al-O-Si linkages facilitate crystallization into mullite, while the slightly less well-mixed prehydrolyzed TEOS sample has silica and alumina domains which are large enough that the silica can begin to crystallize separately from the alumina. In contrast, no mullite was detected in any of the mixed sol samples at any of the temperatures studied. Along a "continuum of mixing," the double alkoxide is thus the best mixed, followed by the prehydrolyzed TEOS sample, and then the samples prepared from the mixed sols. Further evidence of this continuum can be found in the DRIFT and 1-butene data. We have shown previously[19] that increases in the silica stretch frequencies at 900-1330 cm^{-1} correlate with increases in the silica particle or domain size. Others have shown that increases in the frequencies of the Si-O-Si stretching vibrations can be related to a strengthening of the silica network through cross-linking.[22-24] Results of de-convoluting the silica structural peak into its component transverse and longitudinal modes of the asymmetric Si-O-Si stretch are depicted in Figure 9. The relationship between silica domain size and the stretch frequencies holds. The double alkoxide, with its "built-in" Al-O-Si linkages, has the lowest (weakest) silica stretch frequencies, indicating that the silica domains are small. The prehydrolyzed TEOS sample, which is less well-mixed, has slightly higher silica stretch frequencies, but still the silica domains would be expected to be smaller than those found in the mixed sol samples. Finally, the three mixed sol samples have the highest silica stretch frequencies, and are ordered according to the size of the silica particles found in the precursor sols.

Table IV lists the catalytic and acidic properties of these 1 Al:1 Si alumina-silica mixed oxides. Catalytic activity is a function of the homogeneity of the samples, with a maximum 1-butene isomerization activity found in the prehydrolyzed TEOS sample. The double alkoxide-

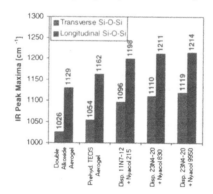

Figure 9. Effect of silica domain size on Si-O-Si bond stretch frequencies.

Table IV. Catalytic and acidic properties of 1 Al: 1 Si alumina-silica mixed oxides

	Double Alkoxide Aerogel	Prehyd. TEOS Aerogel	Dispal 23N4-20 + Nyacol 830	Dispal 23N4-20 + Nyacol 9950	Dispal 11N7-12 + Nyacol 215
1-Butene Isomerization Activity					
(mmol/m^2/h)	0.02	0.44	0.007	0.002	n/a
(mmol/g/h)	12.0	93.0	1.6	0.3	n/a
Lewis/Brønsted Acid Site Ratio	2.6	0.9	2.6	6.8	Only Lewis sites detected

derived sample, which, as a precursor to mullite, does not generate as many Brønsted acid sites as the prehydrolyzed TEOS sample, but its activity is still three times that of the best of the mixed sol samples. Of the mixed sol samples, the Dispal 23N4-20 + Nyacol 830 sample has the highest activity and the lowest Lewis/Brønsted acid site ratio, another indication that it is the "best" mixed of the three.

CONCLUSIONS

Commercial sols, while not ideal model systems, can successfully be used as precursors to prepare alumina-silica mixed oxides at a wide range of compositions. However, a systematic approach is necessary in order to relate the properties of the mixed oxides back to the sol precursors. By first examining silica and alumina as single oxides, we laid the groundwork for an understanding of the mixed oxides prepared from sols. The simple single oxide models based on sol particle size and structure have been extended to explain the properties of the mixed oxide samples. The effects of homogeneity can be framed by a concept of a "continuum of mixing," which is applicable to a wide range of precursor sizes, from monomeric alkoxides to colloidal particles of hundreds of nanometers. This study furthers our aim of providing a foundation for using commercially available sols in the preparation of materials of catalytic interest.

ACKNOWLEDGEMENTS

We thank the National Science Foundation for supporting this work via grant CTS-9522066 and a graduate fellowship for SJM. We also thank Eka Nobel, Inc. and Vista Chemicals, Inc. for providing us with sol samples.

REFERENCES

(1)   Li, D. X.; Thomson, W. J. *J. Mater. Res.* **1990**, *5*, 1963.
(2)   Hyatt, M. J.; Bansal, N. P. *J. Mat. Sci.* **1990**, *25*, 2815.
(3)   Jaymes, I.; Douy, A. *J. Am. Ceram. Soc.* **1992**, *75*, 3154.
(4)   Chu, L.; Anderson, M. A. *Prepr. - Am. Chem. Soc., Div. Pet. Chem.* **1995**, *40*, 84.
(5)   Komarneni, S.; Rutiser, C. *J. Eur. Ceram. Soc.* **1996**, *16*, 143.
(6)   Hoffman, D. W.; Roy, R.; Komarneni, S. *J. Am. Ceram. Soc.* **1984**, *67*, 468.
(7)   Komarneni, S.; Suwa, Y.; Roy, R. *J. Am. Ceram. Soc.* **1986**, *69*, C-155.
(8)   Ismail, M. G. M. U.; Nakai, Z.; Somiya, S. *J. Am. Ceram. Soc.* **1987**, *70*, C-7.
(9)   Wei, W.; Halloran, J. W. *J. Am. Ceram. Soc.* **1988**, *71*, 166.
(10)   Sonuparlak, B. *Adv. Ceram. Mat.* **1988**, *3*, 263.
(11)   Li, D. X.; Thomson, W. J. *J. Am. Ceram. Soc.* **1991**, *74*, 2382.
(12)   Lee, J. S.; Yu, S. C. *J. Mat. Sci.* **1992**, *27*, 5203.
(13)   Fahrenholtz, W. G.; Smith, D. M. *J. Am. Ceram. Soc.* **1993**, *76*, 433.
(14)   Ha, J.; Chawla, K. K. *Ceram. Int.* **1993**, *19*, 299.
(15)   Pach, L.; Iratni, A.; Hrabe, Z.; Svetík, S.; Komarneni, S. *J. Mat. Sci.* **1995**, *30*, 5490.
(16)   Pack, L.; Iratni, A.; Kovar, V.; Mankos, P.; Komarneni, S. *J. Eur. Ceram. Soc.* **1996**, *16*, 561.
(17)   Kara, F.; Little, J. A. *J. Eur. Ceram. Soc.* **1996**, *16*, 627.
(18)   Sheng, G.; Chu, L.; Zeltner, W. A.; Anderson, M. A. *J. Non-Cryst. Solids.* **1992**, *147/148*, 548.
(19)   Monaco, S. J.; Ko, E. I. *Chem. Mater.* Accepted for publication.
(20)   Miller, J. B.; Tabone, E. R.; Ko, E. I. *Langmuir* **1996**, *12*, 2878.
(21)   Basila, M. R.; Kantner, T. R. *J. Phys. Chem.* **1966**, *70*, 1681.
(22)   Miller, J. B.; Johnston, S. T.; Ko, E. I. *J. Catal.* **1994**, *150*, 311.
(23)   Brinker, C. J.; Scherer, G. W. *Sol-Gel Science*; Academic Press: New York, 1990.
(24)   Ying, J. Y.; Benziger, J. B. *J. Am. Ceram. Soc.* **1993**, *76*, 2561.

SOL-GEL PRECURSORS AND THE OXYGEN STORAGE CAPACITY OF PrO_y-ZrO_2 MATERIALS

C.K. NARULA[1*], K.L. TAYLOR[1], L.P. HAACK[1], L.F. ALLARD[2], A. DATYE[3], M.YU. SINEV[4], M. SHELEF[5], R.W. McCABE[5], W. CHUN[5], and G.W. GRAHAM[5*]
cnarula@ford.com, ggraham3@ford.com
1 Chemistry Department, Ford Motor Co., MD 3083, P.O. Box 2053, Dearborn, MI 48121
2 HTML, Oak Ridge National Laboratory, Oak Ridge, TN 37831
3 Department of Chemical and Nuclear Engineering, University of New Mexico, Albuquerque, NM 87131
4 Semenov Institute of Chemical Physics, Russian Academy of Sciences, 4 Kosygin Street, Moscow 11334, Russia
5 Chemical Engineering Department, Ford Motor Co., MD 3179, P.O. Box 2053, Dearborn, MI 48121

ABSTRACT

The gels derived from mixtures of $Pr(O^iC_3H_7)_3$ and $Zr(O^iC_3H_7)_4 \bullet {}^iC_3H_7OH$, upon hydrolysis and pyrolysis, furnish single-phase PrO_y-ZrO_2 materials crystallized in the fluorite structure. These materials can be coated onto high-surface-area γ-alumina powders or deposited onto dense α-alumina coupons in the form of thin films from a solution of parent alkoxides modified with 2,4-pentanedione in THF. The fluorite structure of the PrO_y-ZrO_2 in the films appears to be thermally stable in air up to 1200°C. Temperature-programmed-reduction (TPR) measurements show that the bulk PrO_y-ZrO_2 material with a Pr:Zr molar ratio of 1:1 can store and release oxygen while that with a molar ratio of 1:3 cannot.

The precursors play an important role in determining phase composition of the resulting PrO_y-ZrO_2 material. A mixture of monoclinic and cubic or tetragonal phases was found in PrO_y-ZrO_2 prepared from a new single-source heterometallic alkoxide, $Pr_2Zr_6(\mu_4\text{-}O)_2(\mu\text{-}OAc)_6(\mu\text{-}O^iPr)_{10}(O^iPr)_{10}$, whereas a mixture of cubic and tetragonal phases was present in PrO_y-ZrO_2 made previously by coprecipitation from aqueous solutions of the metal nitrates.

INTRODUCTION

Control of air-to-fuel ratio in an automobile by the standard oxygen sensor results in slight rich/lean excursions of the exhaust gas. Larger transients can occur in response to acceleration/deceleration. Since catalysts used for controlling exhaust gas emissions operate effectively only at a stoichiometric ratio of reducing to oxidizing gases, cerium oxide, an oxygen storage material, is added to the catalyst to damp out such variations [1]. Improved oxygen storage materials are needed, however, due to the demand for increasingly-better catalyst performance and durability. Praseodymium oxide is one of the materials under investigation because it undergoes oxygen exchange at a lower temperature than cerium oxide without significant loss of oxygen storage capacity after high temperature sintering [2]. Due to the difficulties in preparing praseodymia-based materials which allow PrO_2-$PrO_{1.5}$ interconversion, these are not yet commercialized.

Here, we describe the preparation of PrO_y dispersed in zirconia by sol-gel processes employing a mixture of alkoxide precursors and a new single source alkoxide, $Pr_2Zr_6(\mu_4\text{-}O)_2(\mu\text{-}OAc)_6(\mu\text{-}O^iPr)_{10}(O^iPr)_{10}$, showing that the choice of precursor and processing method determines the distribution of various phases of PrO_y-matrix materials and, consequently, the oxygen storage capacity as determined by TPR. We also examine the thermal stability of these materials.

EXPERIMENTAL

The alkoxides were handled carefully in an inert atmosphere in a dry box. Solvents were carefully dried and distilled before use. The alkoxides of cerium, praseodymium and zirconium were prepared by known methods [3]. For calcination, the samples were loaded in a ceramic crucible which was placed in a muffle furnace in an ambient atmosphere. Elemental analyses were carried out at Galbraith Laboratories, Knoxville, TN. Infra-red spectra of these samples were recorded on a Perkin-Elmer system 2000 instrument. Powder X-ray diffraction (XRD) was performed using a Scintag X1 diffractometer. X-ray photoelectron spectroscopy (XPS) with monochromatic Al K_α x-ray radiation was performed using a Surface Science Instruments M-Probe spectrometer.

Preparation of gel from a Mixture of $Pr(O^iC_3H_7)_3$ and $Zr(O^iC_3H_7)_4\bullet^iC_3H_7OH$: A 100 mL flask was charged with $Pr(O^iC_3H_7)_3$ (2.05 g) and $Zr(O^iC_3H_7)_4\bullet^iC_3H_7OH$ (2.5 g) dissolved in THF (40 mL). The reaction mixture was cooled to -78°C and reacted with water (1.0 mL) mixed with THF (20 mL). The gel, thus obtained, was dried and pyrolyzed in air at 600°C (10°C/min, 4h hold) to obtain 1.97 g residue with a Pr:Zr molar ratio of 1:1. Material with a Pr:Zr molar ratio of 1:3 was prepared by appropriately adjusting the ratio of parent alkoxides.

Preparation of PrO_y-ZrO_2 Coated Alumina Particles: A solution of $Pr(O^iC_3H_7)_3$ (0.041 g) and $Zr(O^iC_3H_7)_4\bullet^iC_3H_7OH$ (0.15 g) in THF was added to 0.93 g γ-alumina powder (with ~ 100 m^2 of surface area). The reaction mixture was stirred for two hours and volatiles were removed in vacuum. A portion of residue (0.4 g) was pyrolyzed at 600°C (10°C/min, 4h hold) to obtain PrO_y-ZrO_2 coated particles with a Pr:Zr molar ratio of 1:3. Corresponding material with a Pr:Zr molar ratio of 1:1 was prepared by appropriately adjusting the ratio of parent alkoxides.

Preparation of PrO_y-ZrO_2 films on Alumina Coupons: Clear sols (0.1 M in THF) were prepared by adding 2,4-pentanedione to solutions of $Pr(O^iC_3H_7)_3$ and $Zr(O^iC_3H_7)_4\bullet^iC_3H_7OH$ in THF, cooling the reaction mixture to -78°C, and slowly adding water mixed with THF while stirring. The sols were then warmed to room temperature and stored under a nitrogen atmosphere. A solution with a Pr:Zr molar ratio of 1:1 was obtained by combining the pure sols. PrO_y-ZrO_2 films on α-alumina coupons were prepared by dip coating at a withdrawal rate of 0.4 cm/s, drying to obtain xerogel films, and calcining for 30 minutes at 750°C. The dip-coating/drying/calcination procedure was performed twice in order to double the film thickness. Additional calcination treatments were performed stepwise in a small CMS high-temperature furnace, heating and cooling at a rate of 20°C/minute, and holding at temperature for a period of 30 minutes. The steps in temperature were 900, 1050, 1200, 1300, and 1400°C.

Preparation of $Pr_2Zr_6(\mu_4-O)_2(\mu-O_2CCH_3)_6(\mu-O^iC_3H_7)_{10}(O^iC_3H_7)_{10}$: A 100 ml flask was charged with anhydrous praseodymium acetate (2.12 g, 6.88 mmol), zirconium isopropoxide isopropanolate (9.32 g, 24.03 mmol), and xylene (60 ml). The reaction mixture was heated under reflux for 24 hours, cooled to room temperature, and filtered to remove unreacted praseodymium acetate. The volatiles were removed in vacuo and the residue was dissolved in 13 ml toluene. The solution was stored at -20°C overnight and crystals were isolated by removing supernatant liquid with a syringe. The elemental composition calculated for $Pr_2Zr_6(\mu_4-O)_2(\mu-O_2CCH_3)_6(\mu-O^iC_3H_7)_{10}(O^iC_3H_7)_{10}$ was: C 36.0, H 6.64, Pr 11.75, Zr 22.84; that found was: C 37.5, H 7.0, Pr 11.4, Zr 22.7.

Hydrolysis of $Pr_2Zr_6(\mu_4-O)_2(\mu-O_2CCH_3)_6(\mu-O^iC_3H_7)_{10}(O^iC_3H_7)_{10}$: For hydrolysis, $Pr_2Zr_6(\mu_4-O)_2(\mu-O_2CCH_3)_6(\mu-O^iC_3H_7)_{10}(O^iC_3H_7)_{10}$ (1.05 g, 0.43 mmol) was dissolved in THF (30 mL) and reacted with 2,4-pentanedione (0.26 g, 5.27 mmol). The reaction mixture was cooled to -78° C and hydrolyzed with water (0.095g, 5.27 mmol) mixed with THF (30 mL). The reaction mixture was warmed to room temperature and volatiles were removed in vacuum. The residue was pyrolyzed in air at 600°C (10°C/min, 4h hold) to a light green powder.

RESULTS

The equilibrium phase diagram of praseodymium oxide - zirconium oxide shows that at a concentration of praseodymium oxide below 40%, monoclinic and cubic phases are observed [4]. This mixture of phases transforms to tetragonal and cubic phases in the 1000-1500°C range and to a fluorite-structured phase in the 1500-2200°C range depending on the concentration of PrO_y. For concentrations of PrO_y in the 42-56% range, a pyrochlore phase is observed. Above 56%, the material is a mixture of pyrochlore and tetragonal phases. Among these phases, PrO_2-ZrO_2 crystallized in the fluorite structure is expected to be useful for oxygen storage since it should allow PrO_2-$PrO_{1.5}$ interchange. In a recent study, Sinev et al. reported the preparation of PrO_y-ZrO_2 materials by coprecipitation from aqueous solutions of the metal nitrates [5]. The resulting materials crystallized in a mixture of cubic and tetragonal phases. The concentration of the cubic fluorite-structured phase depended on the concentration of ZrO_2 in the system. In our work, we found that the gels derived from a mixture of $Pr(O^iC_3H_7)_3$ and $Zr(O^iC_3H_7)_4 \cdot {}^iC_3H_7OH$ in 1:3 and 1:1 molar ratios furnish green powders on pyrolysis at 600°C, both of which which are single-phase fluorite-structured materials (Fig. 1).

$$Pr(O^iC3H7)3 + Zr(O^iC3H7)4.^iC3H7OH \xrightarrow{\text{H2O}} \text{GEL}$$

$$\downarrow \text{DRY}$$

$$PrO_y\text{-}ZrO_2 \text{ POWDER} \xleftarrow{\text{600}^0\text{C}} \text{XEROGEL}$$

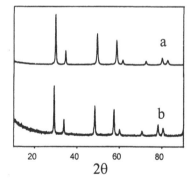

Fig. 1. XRD of PrO_y-ZrO_2 materials derived from $Pr(O^iC_3H_7)_3$ and $Zr(O^iC_3H_7)_4 \cdot {}^iC_3H_7OH$ in (a) 1:3 molar ratio (b) 1:1 molar ratio.

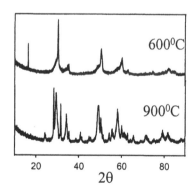

Fig. 2. XRD of PrO_y-ZrO_2 materials derived from single source precursor.

The surface area of the 1:1 material, 120 m^2/g, was reduced to 20 m^2/g after calcining at 800°C. The powder diffraction patterns from both materials did not change with calcination up to 900°C. TPR of the 1:1 material after calcining at 600°C, 800°C, or 900°C shows one peak at 425°C with integrated area corresponding to ~ 400 μmole H$_2$/g. In comparison, the CeO$_2$-ZrO$_2$ system typically shows a peak in the range 550-600°C [6]. These results thus demonstrate that the PrO$_y$-ZrO$_2$ system can release oxygen at a lower temperature than the CeO$_2$-ZrO$_2$ system. TPR of the 1:3 material showed only a small, broad peak above 600°C.

The XPS results reflect this difference in distribution of Pr^{+3} and Pr^{+4} between the samples (Fig. 3). The Pr 3d spectrum of the 1:3 material is characteristic of pure Pr^{+3} while that of the 1:1 material reveals a small amount of Pr^{+4}, indicated by the weak peak at 963 eV (uncorrected binding energy scale). The spectrum of Pr$_{0.55}$Ce$_{0.45}$O$_2$ provides a reference in which the amount of Pr^{+4} is maximal [5].

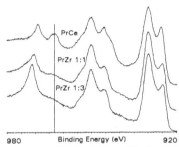

PrCe

PrZr 1:1

PrZr 1:3

980   Binding Energy (eV)   920

Fig. 3. Pr 3d spectra of PrO$_y$-ZrO$_2$ and PrO$_2$-CeO$_2$ powders.

Characterization of PrO$_y$-ZrO$_2$ coated alumina was performed in order to assess the possible effect of ZrO$_2$ on the facile reaction between PrO$_y$ and alumina which is used in most catalyst formulations. The electron micrograph (Fig 4a) shows representative particles of γ-alumina impregnated with a solution of alkoxides having a Pr:Zr molar ratio of 1:3.

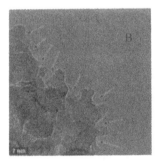

Fig. 4. High resolution electron micrograph of PrO$_y$-ZrO$_2$ coated γ-alumina particles (a). The arrows show damage to the particle during EDS measurements (b).

The EDS analysis of the particle surface clearly showed peaks due to praseodymium and zirconium suggesting that PrO$_y$-ZrO$_2$ forms a coating on the alumina particles. EDS analysis of the interior regions showed very little Pr and Zr suggesting that these elements are present only as a surface layer. Powder XRD patterns obtained from PrO_y-ZrO_2 coated γ-alumina with both 1:3 and 1:1 Pr:Zr molar ratios revealed the presence of crystalline fluorite-structured phases, with the expected lattice constants, in addition to γ-alumina. In TPR measurements of the 1:1 sample, a small peak was found at 425°C with integrated area corresponding roughly to the nominal dilution of bulk 1:1 material with alumina (7wt%).

Thermal evolution of thin (of order 10 nm) films made by dip-coating planar substrates of α-alumina from a solution of the the mixture of single praseodymium and zirconium precursors was followed by XRD. Dense, polycrystalline alumina was used in order to minimize absorption of the solution and also provide a best-case situation with respect to limiting subsequent reaction between the films and the alumina.

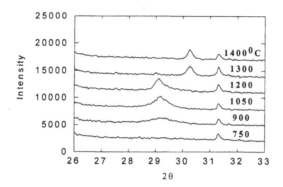

Fig. 5. Changes in XRD pattern of PrO_y-ZrO_2 film on α-alumina coupon upon calcination.

As shown in Fig. 5, films which are nearly invisible (due to low crystallinity) after calcination at 750°C develop a (111) line characteristic of the PrO_y-ZrO_2 fluorite-structured phase by 900°C. This phase is stable up to almost 1200°C, where separation into separate praseodymia and zirconia phases begins to occur. By 1300°C, only the high-temperature-stable tetragonal zirconia phase remains, praseodymia having apparently dissolved into the alumina substrate. (The feature at 31.3° originates from the substrate.)

In another approach to promote the bonding of praseodymium to zirconium during impregnation of alumina, we attempted to employ heterometallic alkoxides. A known heterometallic alkoxide $Pr[Zr(O^iC_3H_7)_5]_3$ is not practical because its preparation requires storage of a mixture of parent alkoxides for several months [7]. We prepared a new precursor from the reaction of praseodymium acetate with zirconium 2-propoxide according to the following reaction.

$$Pr(OOCCH_3)_3 + 3Zr(O^iC_3H_7)_4 \bullet^iC_3H_7OH \longrightarrow Pr_2Zr_6(\mu_4\text{-}O)_2(\mu\text{-}O_2CCH_3)_6(\mu\text{-}O^iC_3H_7)_{10}(O^iC_3H_7)_{10}$$

The elemental analysis of $Pr_2Zr_6(\mu_4\text{-}O)_2(\mu\text{-}O_2CCH_3)_6(\mu\text{-}O^iC_3H_7)_{10}(O^iC_3H_7)_{10}$ supports the formula, and infrared spectra show $\nu_{as}CO_2$ for acetate groups at 1658 and 1572 cm^{-1} shifted to

19

higher frequency with respect to $Pr(OCCH_3)_3$ (1599, 1536 cm^{-1}). The structure is tentatively assigned based on the structure of gadolinium analogue determined by single crystal x-ray crystallography [8]. The pyrolysis at 600°C in air of hydrolyzed and dried precursor leads to a green powder. The XRD pattern (Fig. 2) of this powder shows the crystallization of cubic or tetragonal phases. Further calcination at 900°C furnished a powder with a mixture of monoclinic and cubic or tetragonal phases. The oxygen storage capacity of this material was found to be negligible.

CONCLUSIONS

In conclusion, we have shown that metastable single-phase PrO_y-ZrO_2 material crystallized in the fluorite structure can be prepared by a sol-gel process employing a mixture of $Pr(O^iC_3H_7)_3$ and $Zr(O^iC_3H_7)_4 \bullet {^i}C_3H_7OH$ precursors. The material with 1:1 molar ratio reversibly stores and releases oxygen at 425°C after calcining at 900°C. Further, it is much more stable against reaction with alumina than praseodymia. The thermal stability and ability to release oxygen at a lower temperature than CeO_2-ZrO_2 materials makes it attractive for automotive catalyst applications.

Neither mixed precursors in 1:3 molar ratio nor single source precursor, $Pr_2Zr_6(\mu_4\text{-}O)_2(\mu\text{-}O_2CCH_3)_6(\mu\text{-}O^iC_3H_7)_{10}(O^iC_3H_7)_{10}$, is able to furnish PrO_y-ZrO_2 materials with much oxygen storage capacity.

REFERENCES

1. C.K. Narula, J.E. Allison, D.R. Bauer and H.S. Gandhi, Chem. Mater. **8**, p. 984 (1996).
2. A.D. Logan and M. Shelef, J. Mater. Res. **9**, p. 468 (1994).
3. D.C. Bradley, R.C. Mehrotra and D.P. Gaur, <u>Metal Alkoxides</u>, Academic Press, London, 1978.
4. M.D. Krasil'nikov, I.V. Vinokurov, S.D. Nikitina, Fiz. Khim. Electrokhim. Rasplavl. Tverd. Electrolitov, Tezisy Dokl. Vses. Konf. Fiz. Khim. Ionnykh Rasplanov Tverd. Electrolitov **3**, p. 123 (1979).
5. M.Yu. Sinev, G.W. Graham, L.P. Haack, and M. Shelef, J. Mater. Res. **11**, p. 1960 (1996).
6. A. Trovarelli, R. Zamar, J. Llorca, C. de Leitenburg, G. Dolcetti, and J.T. Kiss, J. Catal. **169**, p. 490 (1997).
7. N.Ya. Turova, N.I. Kozlova and A.V. Novoselova, Russ. J. Inorg. Chem. **25**, p. 1788 (1980).
8. S. Daneiele, L.G. Hubert-Pfalzgraf, J.C. Daran and R.A. Toscano, Polyhedron **12**, p.2091 (1993).

STRUCTURE-PROPERTY RELATIONSHIPS OF BaCeO PEROVSKITES FOR THE OXIDATIVE DEHYDROGENATION OF ALKANES

T. M. Nenoff, N. B. Jackson, J. E. Miller, A. G. Sault, D. Trudell
Sandia National Laboratories, Catalysis and Chemical Technologies Department,
PO Box 5800, Albuquerque, NM 87185-0710

ABSTRACT

The oxidative dehydrogenation (ODH) reactions for the formation of two important organic feedstocks ethylene and propylene are of great interest because of the potential in capital and energy savings associated with these reactions. Theoretically, ODH can achieve high conversions of the starting materials (ethane and propane) at lower temperatures than conventional dehydrogenation reactions. The important focus in our study of ODH catalysts is the development of a structure-property relationship for catalyst with respect to selectivity, so as to avoid the more thermodynamically favorable combustion reaction. Catalysts for the ODH reaction generally consist of mixed metal oxides. Since for the most selective catalyst lattice oxygen is known to participate in the reaction, catalysts are sought with surface oxygen atoms that are labile enough to perform dehydrogenation, but not so plentiful or weakly bound as to promote complete combustion. Also, catalysts must be able to replenish surface oxygen by transport from the bulk.

Perovskite materials are candidates to fulfill these requirements. We are studying $BaCeO_3$ perovskites doped with elements such as Ca, Mg, and Sr. During the ODH of the alkanes at high temperatures, the perovskite structure is not retained and a mixture of carbonates and oxides is formed, as revealed by XRD. While the Ca doped materials showed enhanced total combustion activity below 600°C, they only showed enhanced alkene production at 700°C. Bulk structural and surface changes, as monitored by powder X-ray diffraction, and X-ray photoelectron spectroscopy are being correlated with activity in order to understand the factors affecting catalyst performance, and to modify catalyst formulations to improve conversion and selectivity.

INTRODUCTION

Ethylene and propylene are now the two most important organic feedstocks manufactured by U.S. industry.[1] These chemicals are currently manufactured by steam cracking of natural gas liquids, an extremely energy intensive, nonselective process. The development of catalytic technologies, such as selective oxidative dehydrogenation, with the potential for greater selectivity and lower energy consumption than steam cracking, can have a profound effect on the competitiveness of the U.S. chemical industry, and prevent the movement of these key energy industries to offshore sites.

A general equation describing the oxidative dehydrogenation (ODH) reaction is $C_nH_{2n+2} + O_2 \rightarrow C_nH_{2n} + H_2O$. Because this is an exothermic reaction, the potential exists for low temperature operations compared to straight dehydrogenation methods which are endothermic and challenged by equilibrium constraints. However, the ODH reaction also

21

favors complete oxidation of the alkane. To combat this thermodynamic trend, a catalyst needs to have good selectivity. For selective catalysts the reaction usually involves the lattice oxygen of the material, as shown in the following general formulas: (L = lattice, V = vacancies) (1)$C_nH_{2n} + O_L \rightarrow C_nH_{2n+1} + OH$; (2) $C_nH_{2n+1} + OH \rightarrow C_nH_{2n} + H_2O + O_V$; (3)$O_2 + 2O_V \rightarrow 2O_L$. According to Mars-van Krevelen mechanism[2], oxygen does not necessarily readsorb at the same site that the hydrocarbon reacts. The active catalytic site of the metal oxide is reduced as the alkane is oxidized. This site is then reoxidized by oxygen from a second site in the metal oxide lattice, usually associated with a different type of metal atom. The second site is then reoxidized by oxygen (introduced from the gas phase) and transported through the lattice.

Literature has shown that rare earth metal oxides are of considerable interest for the oxidative dehydrogenation of alkanes.[3] As early as 1978, Mo/V/Nb/O systems[4] have been reported to be very active for ethane oxidative dehydrogenation. However, the catalytically active site is still not well characterized. Perovskites containing rare earth metal oxides possess the unique capability of sustaining nonstoichiometric compositions without affecting their structural integrity. Oxygen-deficient perovskites exhibit ionic conductivity and catalytic activity due to the facile loss and gain of oxygen.[5-11] Oxygen vacancies can be introduced into a rare earth containing perovskite, such as $BaCeO_3$, by doping in metal oxides, so that the new phase exhibits mixed conduction of mobile oxygen ions and electron holes.[11-13] This mobility led researchers to studies which show $SrCe_{1-x}Yb_xO_{3-x}$ (x = 0 - 0.5) as an active catalyst with respect to oxidative dehydrogenation of ethane.[14] Over 400°C, $La_{1-x}Sr_xFeO_{3-\delta}$ is another good perovskite-like ODH catalysts for ethane.[15]

We have chosen to investigate crystalline perovskite-like catalysts with the general formulas $BaCeO_{3-\alpha}$, $BaM_xCe_{1-x}O_{3-\alpha}$, and $Ba_{1-x}M_xCeO_{3-\alpha}$. Perovskites can be formed from reducible oxides, which are required to provide lattice oxygen. Furthermore, they allow for a wide range of stoichiometric substitutions without loss of crystallographic structure, which is easily monitored. These catalysts are being extensively characterized by powder X-ray diffraction (XRD), and X-ray Photoelectron Spectroscopy (XPS) and were tested for catalytic activity (as monitored by gas chromatography (GC)) with respect to the oxidative dehydrogenation of ethane to ethylene.

EXPERIMENT
Catalyst Preparation:
Our catalysts were the following target perovskite-like compounds: $BaCeO_{3-\alpha}$, $BaM_xCe_{1-x}O_{3-\alpha}$, $Ba_{1-x}M_xCeO_{1-\alpha}$, with x = 0.1-0.5. The starting materials (all from Sigma Chemicals) were $Ba(CO_3)_2$ (99.999%), CeO_2 (99.999%) and dopant metals (La_2O_3, 99.999%; Y_2O_3, 99.999%; Nb_2O_3, 99.999%; MgO, 99.99%; $Ca(CO_3)_2$, 99.999%; $Sr(CO_3)_2$, 99.995%). They were mixed for three hours in an agate ball mill. Each mixture was heated (300°C/hour ramp rate) in air at 1250°C for 5 hours in platinum crucibles. Each sample was milled, and then refired in the crucibles under the previous conditions.

Note that alkaline earth samples were pressed into pellets (40-60+ mesh size) before final firing and retained their pellet forms. All starting materials and catalyst products were stored under inert atmosphere (N_2) until needed for reactions or testing.

Bulk Catalyst Characterization:

Powder **X-ray** diffraction data were collected at room-temperature on a Siemens Model D500 automated diffractometer, with Θ-2Θ sample geometry and Cu K_α radiation, between $2\Theta = 5$ and $60°$, step size $0.05°$.

ODH Catalyst Testing:

The catalysts were tested for ODH of ethane in a stainless steel flow system. Samples were tested either in powder or pellet form (40-60+ mesh) by flowing 83 sccm of a 10.5% O_2 in N_2 mixture and 17 sccm C_2H_6 over 0.250 g of material in a 4.9 mm I.D. stainless steel tube. The reaction temperatures, measured with a thermocouple positioned inside the reactor tube just above the catalyst, ranged from 350 to 750°C. Reactants and products were analyzed with an on-line MTI GC utilizing a 10 m molecular sieve 5A PLOT column (O_2, CH_4, CO) and a 8 m Poraplot Q column (CO_2, C_2H_4, C_2H_6). No additional products other than water were detected. Tests conducted in an empty reactor tube indicated that the thermal contribution to the reaction was negligible below 550 °C, increasing to 2% propane conversion at 600 °C, 5% at 650°C, 30% at 700 °C, and 40% at 750 °C.

Catalyst Surface Characterization:

X-ray Photoelectron Spectroscopy (XPS) measurements were made in a combined ultra-high vacuum (UHV) surface analysis/atmospheric pressure reactor system[16,17] that allows measurement of the surface properties of catalytic materials before and after exposure to reactive environments without intervening exposure to air. With this system, various perovskite catalysts were analyzed following treatment in reactive environments to simulate reaction conditions. The treatments were designed to simulate increasing conversion of ethane and consumption of oxygen with increasing reaction temperature. Thus, the conditions studied included treatment in a stoichiometric mixture of 108 Torr ethane and 54 Torr oxygen at 500°C for 1 h, treatment in lean mixture of 108 Torr ethane and 10 Torr oxygen at 600°C for 1h, treatment in 108 Torr ethane at 700°C for 1h, and treatment in 108 Torr ethane plus 54 Torr CO_2 at 700°C for 1h. Prior to all treatments the samples were outgassed in vacuum for five minutes at the desired reaction temperature to remove any volatile organic species. XP spectra were taken following each treatment using a non-monochromatic Mg $K\alpha$ source with an analyzer resolution of 1.0 eV. Because of the insulating nature of the perovskites, some sample charging always occurred and binding energies were referenced to the Ba $3d_{5/2}$ peak at 779.7 eV.[18,19] Spectra were collected in the Ba 3d, Ce 3d, O 1s, and C 1s regions for each sample, as well as the Ca 2p or Mg 2p regions for samples containing these modifiers. All peaks were subjected to Tougaard background subtraction followed by integration of peak areas to determine relative XPS signal intensities.

RESULTS and DISCUSSION

Our motivation for studying metal oxide perovskite-like materials as oxidative dehydrogenation catalysts is two-fold. First, as described above, we hope to enhance utilization of alkanes as precursors to the commodity chemicals (propylene and ethylene) commonly used in industry. Second, we would like to study the factors affecting catalyst performance, such as activity and selectivity. An end result would be the understanding of structure-property relationships of the catalysts, and a predictive capability in choice of catalyst per specific reaction.

Catalyst Testing: $BaCeO_3$ and $BaCe_{1-x}M_xO_{3-\alpha}$ (M = Ca, La, Y, Nd; x = 0.05, 0.10, 0.15, 0.20) phases were tested for the ODH of ethane in the powder form. Each of these materials showed enhanced activity for propane conversion when compared to the thermal reaction at temperatures less than 650 °C. However, CO_2 was the favored product and the ethylene yield was never significantly improved over that of the thermal reaction. At 700°C, many of the materials, most notably the Ca doped materials showed enhanced conversion of ethane to ethylene. For example the Ca0.10 material showed a 34% yield (ethylene produced / ethane fed) of ethylene compared to a 17% yield for the thermal runs. For catalysts that exhibit enhanced ethylene yield, the selectivity to ethylene shows sharp increases in the temperature range of 650-700°C (see figure 1).

XRD analysis of catalysts recovered from the reactor show that the perovskites have significantly degraded, and that new phases have formed (see below, cubic fluorite structures). These observations suggest that at temperatures greater than 650 °C, the ethylene yield was enhanced either through the destruction of a combustion enhancing perovskite phase, through the formation of new more selective phase, or a combination of both. In contrast to this result, cycling the reactor to 750 °C changed the selectivity over the material at lower temperatures. For example, in one experiment the ethylene yield for the Ca0.10 material at 700 °C was reduced by more than half by cycling the reactor to 750 °C. This suggests that the perovskite, or at least an intermediate, phase may be the more selective phase at elevated temperatures.

Drawing on these results several new catalysts were produced. These new materials were based on the $BaCeO_3$ baseline material but were doped with Group II elements (Mg, Ca, and Sr). Since oxygen vacancies may have led to combustion reactions, the precursors were mixed in ratios intended to result in dopant substitution on the divalent Ba site. In addition, to remove pressure drop effects, the materials were pelletized. The materials from this family tested for catalytic activity to date include $Ba_{0.8}M_{0.2}CeO_{3-\alpha}$ (M = Mg, Ca, Sr) and $BaCe_{0.8}Ca_{0.2}O_{3-\alpha}$. None of these materials, including $BaCe_{0.8}Ca_{0.2}O_{3-\alpha}$ which had performed fairly well in the powder form, yielded as much ethylene as the thermal (non-catalytic) run at 700 °C. In part this was probably

24

due to a decrease in exposed catalytic surface area. An interesting phenomenon was noted when the activity was monitored over several hours time. The yield (and selectivity) of ethylene just after the reactor had stabilized at 700 °C was consistently much higher than that fifteen minutes into the reaction (see figure 2). This then reached a minimum value before gradually recovering some, but not all of its selectivity (and hence yield) with a leveling occurring several hours into the reaction.

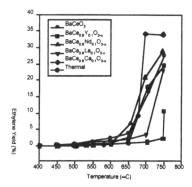 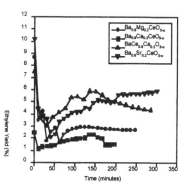

Figure 1: *Ethylene yield versus temperature for perovskite powders.*

Figure 2: *Ethylene yield versus time at 700°C for perovskite granules.*

XRD: The perovskite-like doped barium cerium oxide phases had a moderate hardness and were very dense (pellets were pressed for ease of gas flow during catalytic reactions). The XRD of the pure $BaCeO_3$ and the $BaM_xCe_{1-x}O_{3-\alpha}$ showed those of the tetragonal single phase which have been previously reported in the literature.[20] (The space group is a tetragonal extension of the cubic perovskite unit cell.)

The catalyst samples from figure 2 were characterized by XRD after recovery. XRD data shown in Figure 3 is exemplary of the $BaM_xCe_{1-x}O_{3-\alpha}$ series. The catalyst remains tetragonal through 600°C. However, at 700°C, there is clearly a break down of the original perovskite phase; the resultant phases are a combination of $BaCO_3$, CeO_2, and possibly $Ce_{1-x}M_xO_{2-0.5x}$ (a cubic fluorite structure). The broadness of the XRD peaks can be due to overlapping peaks of the various X values of the fluorite phase and/or a decrease in particle size of the bulk catalyst due to temperature effects.

XPS: Initial XPS analysis of all of the perovskites reveals the expected presence of Ba, Ce, C and O on the surface of the catalyst, as well as Mg or Ca for perovskites containing these dopants. The Ce 3d/Ba $3d_{5/2}$ intensity ratio, after correcting for relative sensitivity factors,[18,19,21] corresponds to the expected Ba/Ce atomic ratio of ~1.0 for all of the perovskites studied. Upon subjecting the perovskites to thermal treatments in various reactive environments, the Ce 3d/Ba $3d_{5/2}$ ratio invariably declines with increasing severity of the treatment, ultimately approaching zero following treatment in ethane/CO_2 mixtures at 700°C (figure 4). At this stage, both the C 1s and O 1s XPS regions reveal the presence of substantial amounts of carbonate, and very little metal oxide. Very little CeO_2 surface is present relative to the $BaCO_3$ surface, and the surface sensitive XPS technique detects mainly Ba. The catalyst surface following reaction at 700°C clearly consists mainly of $BaCO_3$. This is further confirmed by reports on other ODH catalysts.[14] (see below)

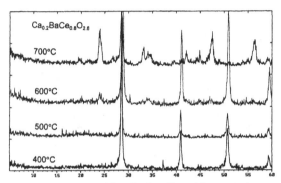

Figure 3*: Powder X-ray diffraction of $Ca_{0.2}BaCe_{0.8}O_{2.8}$ with respect to temperature.*

We focused on one of the better performers, $Ca_{0.2}BaCe_{0.8}O_{2.8}$. The XPS analysis of the Ca containing perovskites indicates that both additives are initially present on the surface in amounts that are close to or slightly below the expected stoichiometric values. Upon subjecting the Ca containing materials to increasingly severe reaction conditions, the Ca 2p signal decreases relative to the Ba $3d_{5/2}$ signal, ultimately falling to a value approximately one third that expected for a uniform stoichiometric mixture. During decomposition of the perovskite structure, the Ca becomes associated primarily with the CeO_2 phase, which exposes very little surface area to the XPS probe. (Phases identified by XRD, see above.)

It should be mentioned that the Ce 3d peak shapes prior to any treatments do not correspond to fully oxidized Ce^{4+}. Instead the spectra indicate a mixture of Ce^{3+} and Ce^{4+}. Upon heating in vacuum, the extent of oxidation of Ce generally increases, indicating that diffusion of oxygen from the bulk to the surface is occurring. During treatment in reactive environments, the Ce becomes more highly reduced, especially for treatments involving lean ethane/oxygen mixture or pure ethane. Taken together these results are consistent with the expectation that these perovskite materials will allow efficient oxygen transport, and that Ce will undergo facile redox chemistry under reaction conditions. Both of these properties are essential to the proper functioning of oxidative dehydrogenation catalysts.

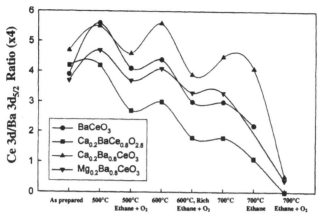

Figure 4: *Decline of Ce (as monitored by XPS) on catalyst surface with temperature.*

Literature[14] reports have suggested that with the ODH catalyst $SrCe_{1-x}Yb_xO_{3-0.5x}$, decomposition occurs above 700°C, while conversion increases and selectivity decreases of the ethane. They show that $SrCO_3$ and $Ce_{1-x}Yb_xO_{2-0.5x}$ are formed. The result is explained as the Ce containing phase having a potentially higher oxygen defect concentration than the parent oxide, and is postulated to be reduced from the 4+ to 3+ oxidation state. Combining this information with our characterization data, we show that at 700°C, there is a great amount of oxygen availability in the catalyst. XPS shows a changing in the oxidation state of the cerium from 4+ to 3+, which eventually leads to a collapse of the perovskite-like structure and resulting in a carbonate and oxide (with variability as to the location of the dopant in the structure). Furthermore, our catalyst testing shows that there is an enhanced activity and selectivity of our alkali metal doped catalysts at 700°C in the initial moments of the reaction for granules and for longer periods with powders. One explanation is that the perovskite-like phase is at peak catalytic performance (activity and selectivity) for the ODH of ethane before collapsing to the cubic fluorite phase. Our intent is to isolate the most active phase and further study its structure property relationships during catalysis.

CONCLUSIONS

In our catalytic studies of doped Ba/Ce/O perovskite-like metal oxides, the structure of the catalyst is lost before 700°C. In-depth characterization of the catalyst by XPS has found that the majority of the working catalytic surface area consists primarily of $BaCO_3$. However, further study is needed to understand if this is due to particle size effects or actual atom migration during reactions. The XPS studies also confirm literature XRD accountings[20] of cerium oxidation state changes during the ODH of ethane in similar materials.

Future work will focus on more fully understanding the effects of dopant elements (oxidation states and atomic radii) on these Ba/Ce/O perovskite-like metal oxides materials. Specifically, we will focus on synthetically mapping the composition range and substitution effects for the alkaline earth metals as dopants, and characterizing with XPS, TEM, GC and quantitative XRD analyses of the catalysts and the resultant metal oxide phases.

ACKNOWLEDGMENTS

This work was supported by the United States Department of Energy under contract DE-AC04-94AL85000. Sandia is a multiprogram laboratory operated by Sandia Corporation, a Lockheed Martin Company, for the United States Department of Energy.

REFERENCES

1. E. S. Linpinsky, J. D. Ingham, Brief Characterization of the Top 50 US Commodity Chemicals, USDOE, ILA 207376-A-H1, Sept. 1994. Also see: Chemical & Engineering News, April 8, 1996, p. 17.

2. J. A. Labinger, K. C. Ott, S. Metha, H. K Rockstad, S. Zoumalan. J. Chem. Soc., Chem. Comm., p. 543 (1987).

3. S.R. Vatcha, Catalytica, Studies Division: Oxidative Dehydrogenation and Alternative Dehydrogenation Processes, Study Number 4192 OD, 1993.

4. E. M. Thorsteinson, T. P. Wilson, F. G. Young, P. H. Kasai, J. Catal. **52**, p. 116 (1978).

5. R. A. Bayerlein, A. J. Jacobson, K. R. Poeppelmeyer, US 4,482,644, 1984.

6. R. A. Bayerlein, A. J. Jacobson, K. R. Poeppelmeyer, US 4,503,166, 1985.

7. R. A. Bayerlein, A. J. Jacobson, K. R. Poeppelmeyer, J. Chem. Soc., Chem. Commun., p. 225 (1988).

8. R. J. H. Verhoeve, Advanced Materials in Catalysis; Burton, J. J.; Garten, R. L., Eds.; Academic: New York, 1977, p. 129.

9. H. Iwahara, T. Esaka, H. Uchida, N. Maeda, Solid State Ionics, 3/4, p. 359 (1981).

10. T. Yajima, H. Iwahara, Solid State Ionics, 50, p. 281 (1992).

11. H. Iwahara, H. Uchida, K. Ono, K. Ogaki, J. Electrochem. Soc. **135(2)**, p. 529 (1988).

12. M. K. Paria, H. S. Maiti, Solid State Ionics. **13**, p. 285 (1984).

13. D. Mastromonaco, I. Barbariol, A. Cocco, Ann. Chim. (Rome). **59**, p. 465 (1969).

14. O. J. Velle, A. Andersen, K.-J. Jens, Catalysis Today. **6**, p. 567 (1990).

15. G. Yi, T. Hayakawa, A. G. Andersen, K. Suzuki, S. Hamakawa, A. P. E. York, M. Shimizu, K. Takehira, Catalysis Letters. **38**, p. 189 (1996).

16. A. G. Sault, E. P. Boespflug, C. H. F. Peden, J. Phys. Chem. **98**, 1652 (1994).

17. A. G. Sault, J. Catal. **156**, 154 (1995).

18. C. D. Wagner, H. A Six, W. T. Jansen, J. A. Taylor, Appl. Surface Sci. **9**, 203 (1981).

19. C. D. Wagner, W. M. Riggs, L. E. Davis, J. F. Moulder, G. E. Muilenberg, Handbook of X-ray Photoelectron Spectroscopy. Perkin-Elmer Corp., Eden Prairie, MN, 1979.

20. V. Longo, F. Ricciardiello, D. Minichelli, J. Materials Chem. **16**, 3503 (1981).

21. C. D. Wagner, L. E. Davis, M. V. Zeller, J. A. Taylor, R. H. Raymond, L. H. Gale, Surface Int. Anal. **3**, 211 (1981).

DENSE OXIDE MEMBRANES FOR OXYGEN SEPARATION AND METHANE CONVERSION

A. J. JACOBSON, S. KIM, A. MEDINA, Y. L.YANG AND B. ABELES,
University of Houston, Department of Chemistry, Houston, TX 77204-5641.

ABSTRACT

Methane conversion to synthesis gas in membrane reactors that use dense mixed electronic-ionic conducting membranes for oxygen separation has received much recent attention. The oxygen flux achievable in these reactors depends on a combination of the bulk diffusion rate for oxygen transport and the surface reaction rate for oxygen activation and either recombination or reaction with methane. Here we compare recent oxygen permeation data for tubular membranes of $La_{0.5}Sr_{0.5}Fe_{0.8}Ga_{0.2}O_{3-\delta}$ with our previous results for $SrCo_{0.8}Fe_{0.2}O_{3-\delta}$, $Sm_{0.5}Sr_{0.5}CoO_{3-\delta}$, $SrFeCo_{0.5}O_{3.25-\delta}$. The pressure dependence of the oxygen permeation flux has been measured at different temperatures and used to determine the relative importance of bulk and surface kinetics for these oxides.

INTRODUCTION

Mixed-conducting oxides with high oxygen ion conductivities can form the basis for ceramic membranes that separate oxygen from air. Such membranes are also of interest for their potential use in membrane reactors that can produce synthesis gas ($CO+H_2$) by direct conversion of hydrocarbons such as methane. For this application, materials are required to have stability in reducing atmospheres, appropriate mechanical properties as well as high oxygen permeability. Electronic conductivity must also be maintained at the low oxygen partial pressures ($\sim 10^{-17}$ atm) on the hydrocarbon side of the membrane.

Since the early results of Teraoka et al. [1, 2] several groups have investigated the use of mixed electronic-ionic conducting perovskite oxides and related materials as membranes for these applications. [1-8] It is now well established that in general the oxygen flux achievable depends on a combination of the bulk diffusion rate for oxygen transport and the surface reaction rate for oxygen activation and either recombination or reaction with methane. The bulk diffusion rate depends on the coupled transport of oxygen vacancies and electron holes both of which depend on the specific metal atom substituents present in a particular oxide system and on the oxygen non-stoichiometry. The latter depends on the temperature and the oxygen partial pressures on either side of the membrane.

The surface reaction rates also depend on the specific oxide composition and the oxygen stoichiometry but the surface kinetics are less well understood than the bulk diffusion. Kilner has shown [9] that the surface exchange coefficients (k) measured by isotope exchange under equilibrium conditions at high oxygen partial pressures (0.2 -1 atm) for both perovskite and fluorite structure oxides correlate with the diffusion coefficients (D) measured under the same conditions. For the fluorite oxides the slope of a plot of logk versus logD is close to 1 whereas for the perovskites the corresponding slope is 0.5. Little information is currently available about perovskite oxide surface reactivity at the low oxygen partial pressures relevant to methane conversion though some data for $LaCr_{1-x}Ni_xO_{3-x}$ were recently reported. [10]

In this paper, we review our previous oxygen permeation results for $SrCo_{0.8}Fe_{0.2}O_{3-\delta}$, $Sm_{0.5}Sr_{0.5}CoO_{3-\delta}$, $SrFeCo_{0.5}O_{3.25-\delta}$ tubular membranes [11,12] together with some recent data on $La_{0.5}Sr_{0.5}Fe_{0.8}Ga_{0.2}O_{3-\delta}$. All of these data were obtained with a small partial pressure gradient across the membrane, typically in the range $1 \leq pO_2 \leq 0.001$ atm. We have developed a model to describe the transport in this partial pressure range and to obtain both the ambipolar diffusion coefficient and the surface exchange rate constant. The results can be compared with the data from

29

isotope exchange experiments. This particular set of compositions are found to span the range of behavior from surface to bulk limited oxygen transport for membranes of comparable thickness.

EXPERIMENTAL

The synthesis of $SrCo_{0.8}Fe_{0.2}O_{3-\delta}$, $Sm_{0.5}Sr_{0.5}CoO_{3-\delta}$, and $SrFeCo_{0.5}O_{3.25-\delta}$ powders and fabrication of tubular membranes have been described in previous publications. [11, 12]

The ceramic powder for the $La_{0.5}Sr_{0.5}Fe_{0.8}Ga_{0.2}O_{3-\delta}$ membrane was produced by freeze-drying nitrate solutions. Pre-dried La_2O_3(Aldrich, 99.99%) and $SrCO_3$ (Aldrich, 99.995%) were dissolved in nitric acid together with Fe metal powder (Aldrich, 99.99+%) and Ga metal (Alfa 99.999%) in the required stoichiometric ratio. The nitric acid solution was then sprayed using an atomizer (Sonotek) into liquid nitrogen at a rate of 2.5 ml/min. The resulting nitrate "snow" was freeze-dried using an FTS Dura-Dry II MP freeze-dryer. The resulting nitrate powder was gradually heated from 90°C to 300°C to remove residual acid and then heated at 800°C for 2h to decompose the nitrates. Finally the powder was fired in an alumina crucible in air at 1340°C for 10h and cooled to room temperature by shutting off the furnace. Characterization of this and subsequent samples by X-ray powder diffraction was carried out with a Scintag XDS 2000 powder diffractometer using CuKα radiation. The stability of $La_{0.5}Sr_{0.5}Fe_{0.8}Ga_{0.2}O_{3-\delta}$ was investigated by thermogravimetric analysis using a TA Instruments 2950 analyzer. Data were obtained at $pO_2 = 1.0$ and 0.01 atm at a heating rate of 2°C/min.

To fabricate a dense tubular membrane the final powder was ball milled for 72 h in ethanol and then mixed with 0.20 wt % stearic acid lubricant (Aldrich, 99+%). This mixture was formed into a tube by cold isostatic pressing at 40,000 psi using a custom-designed rubber mold. The green tube was heated at 0.1 °C/min to 390°C in oxygen to first remove the stearic acid, followed by heating at 0.5°C/min to 1340°C. After 10 h at 1340°C the tube was cooled to room temperature at 0.5°C/min. The tube had a density of >97% of theoretical. The ends of the tube were then cut square and ground on 600 grit SiC paper for final installation into the permeation apparatus.

The details of the preparation of membranes of the other materials and the measurement apparatus for the tubular membranes have been described in previous papers [11].

RESULTS

<u>Sample Characterization</u>   The $La_{0.5}Sr_{0.5}Fe_{0.8}Ga_{0.2}O_{3-\delta}$ sample was shown by X-ray diffraction to be a single phase cubic perovskite. The composition of the as prepared powder was determined to be $3-\delta = 2.88$ by reduction of the sample in hydrogen. The composition changes as a function of temperature at $pO_2 = 1.0$ and 0.01 atm are shown in Fig. 1.

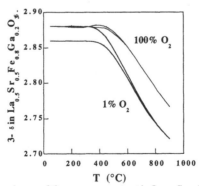

Fig.1: The temperature dependence of the oxygen content in $La_{0.5}Sr_{0.5}Fe_{0.8}Ga_{0.2}O_{3-\delta}$.

At $pO_2 = 1$ atm, the sample begins to reduce at ~ 400°C. The weight change continues until at 900°C the composition 3-δ = 2.77. On cooling, the reoxidation is reversible above 400°C. At this point, the gas composition was adjusted to give $pO_2 = 0.01$ atm. Again reduction is observed above 400°C but the final composition at 900°C is as expected lower and corresponds to 2.72. The final composition on reoxidation is 2.86.

Oxygen Transport kinetics   In reporting the permeation data, we use previously derived relations that provide information about the relative roles of surface exchange versus bulk diffusion in the oxygen transport [11]. Under the assumptions discussed by Kim et $al.$ [11], if the oxygen transport is limited by the rate of oxygen exchange between the gas and the membrane surface the following relation holds:

$$F = \frac{\pi r_1 r_2 w c_i k_{io}}{r_1 + r_2}\left(\sqrt{P_1/P_0} - \sqrt{P_2/P_0}\right) \qquad (1)$$

where r_1 and r_2 are the radial coordinates of the outer and inner tube walls respectively, w is the tube length, c_i is the density of oxygen ions, p_1 and p_2 are the inlet and outlet oxygen pressures p_o is 1 atm oxygen pressure and k_{io} is the surface exchange coefficient at 1 atm pressure. Eq. (1) predicts that under purely surface limited transport conditions a plot of the oxygen flow rate versus the parameter $(p_1/p_o)^{0.5}-(p_2/p_o)^{0.5}$ will be linear

For the alternative case in which transport is limited by the bulk diffusion rate through the membrane the flow rate F is given by:

$$F = \frac{\pi w c_i D_a}{2\ln(r_1/r_2)}\ln(p_1/p_2) \qquad (2)$$

where D_a is the ambipolar oxygen ion-electron hole diffusion coefficient and the flow rate is now linearly proportional to $\ln(p_1/p_2)$.

Permeation Measurements   The measured oxygen flow rates for a tubular membrane of $La_{0.5}Sr_{0.5}Fe_{0.8}Ga_{0.2}O_{3-\delta}$ are plotted in Figs. 2a and 2b against the pressure terms in Eqs. (1) and (2) respectively. The dimensions of the tubular membrane were $r_1 = 0.60$cm, $r_2 = 0.50$c, and $w = 2.94$ cm.

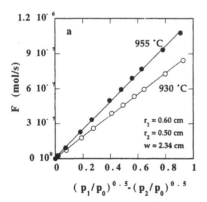 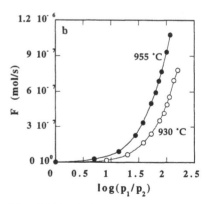

Fig. 2: Oxygen flow plotted vs. (a) $(p_1/p_0)^{0.5}-(p_2/p_0)^{0.5}$ and (b) $\log(p_1/p_2)$ in a tubular membrane with the composition $La_{0.5}Sr_{0.5}Fe_{0.8}Ga_{0.2}O_{3-\delta}$ measured at the indicated temperatures. The inlet pressure p_1 (outside tube) varied from 0.001 to 1 atm.

The linear dependence in Fig. 2a when compared with the non-linear behavior in Fig. 2b demonstrates that transport is surface exchange limited with little or no contribution from the bulk diffusion kinetics. From the slopes of the lines in Fig. 2 and assuming an average value of $c_i = 0.075$ mol/cm^3 based on the thermogravimetric data, the surface exchange rates can be calculated from Eq. (1). Values of 7.35 and 5.61 10^{-6} cm/sec were obtained at 955°C and 930°C, respectively.

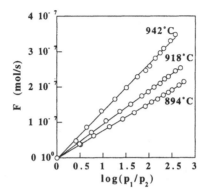

Fig. 3 Oxygen permeation data for a tubular membrane of $SrFeCo_{0.5}O_{3.25-\delta}$. The dimensions of the tube were $r_1 = 0.60$ cm, $r_2 = 0.50$ cm, and $w = 2.9$ cm. Data are from reference 12.

Comparable data are shown in Fig. 3 for a tubular membrane of $SrFeCo_{0.5}O_{3.25-\delta}$ with similar dimensions. It is apparent that while the oxygen flows are comparable, the permeation in this case is limited by the bulk diffusion. Because this membrane is a composite containing mostly $SrFe_{1.08}Co_{0.42}O_{3.25-\delta}$ (70%) with an average composition $SrFeCo_{0.5}O_{3.25-\delta}$ only an effective diffusion coefficient can be derived. Based on the slopes of the plots in Fig. 3, the values of this effective diffusion coefficient were determined from Eq. 2 to be 4.48×10^{-8}, 3.18×10^{-8} and 2.58×10^{-8} cm^2/sec at 942, 918, and 894°C, respectively.

DISCUSSION

In previous work, we have shown that in contrast to the examples shown above, the properties of $SrCo_{0.8}Fe_{0.2}O_{3-\delta}$ and $Sm_{0.5}Sr_{0.5}CoO_{3-\delta}$ membranes are intermediate in behavior. For membranes of comparable thickness, the oxygen permeation flux is limited by both bulk diffusion and surface kinetics and permeation measurements can be used to obtain values of both D_a and k. The ambipolar diffusion coefficients for $SrCo_{0.8}Fe_{0.2}O_{3-\delta}$, and $Sm_{0.5}Sr_{0.5}CoO_{3-\delta}$, are compared in Fig. 4 with the effective diffusion coefficient of the membrane with the overall composition $SrFeCo_{0.5}O_{3.25-\delta}$. The results span two decades with the highest value being observed for $SrCo_{0.8}Fe_{0.2}O_{3-\delta}$. The surface exchange coefficients for $SrCo_{0.8}Fe_{0.2}O_{3-\delta}$ and $Sm_{0.5}Sr_{0.5}CoO_{3-\delta}$ are also compared with the results obtained for $La_{0.5}Sr_{0.5}Fe_{0.8}Ga_{0.2}O_{3-\delta}$. Although the data for the iron gallium sample are limited, the observed k values are substantially lower than observed for the cobalt containing samples.

A correlation between the diffusion coefficients and surface exchange coefficients determined by SIMS analysis of ^{18}O exchange profiles has been reported for both perovskite and fluorite oxide systems. [9] Plots of logk versus logD were shown to have constant slopes for a wide range of compositions with a value of 0.5 for the perovskite oxides. The results that we have obtained for $SrCo_{0.8}Fe_{0.2}O_{3-\delta}$ and $Sm_{0.5}Sr_{0.5}CoO_{3-\delta}$ from permeation measurements closely follow the same correlation. The data are shown in Fig. 5 for the two compounds at different temperatures. A fit to these data gives a slope of 0.47 in excellent agreement with the previous results. The correlation in Fig 5 can be used to estimate a surface exchange coefficient of $\sim 1 \times 10^{-5}$ cmsec^{-1} at 894°C for $SrFeCo_{0.5}O_{3.25-\delta}$ and a diffusion coefficient for $La_{0.5}Sr_{0.5}Fe_{0.8}Ga_{0.2}O_{3-\delta}$ at 955°C of $\sim 1 \times 10^{-8}$ cm^{2}sec^{-1}.

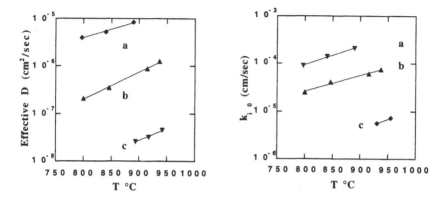

Fig 4 Comparison of the effective ambipolar diffusion coefficients [left] for (a) $SrCo_{0.8}Fe_{0.2}O_{3-\delta}$, (b) $Sm_{0.5}Sr_{0.5}CoO_{3-\delta}$, and (c) $SrFeCo_{0.5}O_{3.25-\delta}$ and surface exchange coefficients [right] for (d) $SrCo_{0.8}Fe_{0.2}O_{3-\delta}$, (e) $Sm_{0.5}Sr_{0.5}CoO_{3-\delta}$ and (f) $La_{0.5}Sr_{0.5}Fe_{0.8}Ga_{0.2}O_{3-\delta}$.

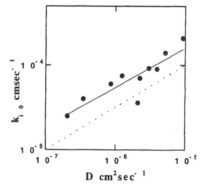

Fig. 5 The experimental diffusion and surface exchange coefficients obtained from permeation measurements for $SrCo_{0.8}Fe_{0.2}O_{3-\delta}$ and $Sm_{0.5}Sr_{0.5}CoO_{3-\delta}$. The dotted line is from ref [9]

CONCLUSIONS

We have shown that both diffusion and surface exchange coefficients can be obtained from permeation measurements as a function of the pressure gradient using a simple model to describe the transport. A similar correlation between D and k is observed as in the case of isotope exchange experiments.

ACKNOWLEDGMENTS

We thank the Texas Center for Superconductivity and the Robert A. Welch Foundation for financial support of this work. The work was supported by the MRSEC program of the National Science Foundation under Award Number DMR-9632667.

REFERENCES

1. Y. Teraoka; T. Nobunaga; N. Yamazoe, Chem. Lett., 503 (1988).

2. Y. Teraoka; H. Zhang; S. Furukawa; N. Yamazoe, Chem. Lett., 1743 (1985).

3. T. J. Mazanec, Proceedings of the First International Conference on Ceramic Membranes, Proc. Electrochem Soc. , eds. H.U. Anderson, A.C. Khandkar and M. Liu, Pennington, NJ, PV95-24, 16 (1997).

4. H. J. M. Bouwmeester; H. Kruidhof; A. J. Burggraaf, Solid State Ionics **72**, 185 (1994).

5. R. H. E. van Doorn; H. Kruidhof; H. J. M. Boumeester; A. J. Burggraaf, Proceedings of the First International Conference on Ceramic Membranes, Proc. Electrochem Soc. eds. H.U. Anderson, A.C. Khandkar and M. Liu, Pennington, NJ, PV95-24, 138 (1997).

6. U. Balachandran; J. T. Dusek; R. L. Mieville; R. B. Poeppel; M. S. Kleefisch; S. Pei; T. P. Kobylinski; C. A. Udovich; A. C. Bose, *Applied Catalysis A: General,* **133**, 19-29, (1995).

7. T. H. Lee, Y. L. Yang, A. J. Jacobson, B. Abeles; M. Zhou, *Solid State Ionics* **100**, 77-85 (1997).

8 T. H. Lee, Y. L. Yang, A. J. Jacobson, B. Abeles; S. Milner, *Solid State Ionics* **100**, 87-94 (1997).

9. J. Kilner, Second International Symposium on Ionic and Mixed Conducting Ceramics, Ed. T. A. Ramanarayanan, W. L. Worrell and H. L. Tuller, Electrochemical Society Proceedings Vol. 94-12, 174 (1994).

10. C. A. Mims; M. Stojanovic; N. Joos; H. Moudallal; A. J. Jacobson, Proc. Electrochem. Soc. PV97-40, 737-745 (1997).

11. S. Kim; Y. L. Yang; A. J. Jacobson; B.Abeles; B., Solid State Ionics in press.

12. S. Kim; Y. L.Yang; R. Christoffersen; A. J.Jacobson, MRS Proceedings accepted 1997.

COMPARISON OF NANOCRYSTALLINE AND POLYCRYSTALLINE OXIDE SUPPORTS FOR CATALYTIC OXIDATION OF METHANE

Horst H. Hahn[1], Heiko Hesemann[1], William S. Epling[2] and Gar B. Hoflund[2*]

[1]Darmstadt University of Technology, FB 21-Materials Science Department, FG Thin Films, Petersenstr. 23, 64287 Darmstadt, Germany

[2]Department of Chemical Engineering, University of Florida, Gainesville, FL 32611, USA, fax: 001-352-392-9513

ABSTRACT

Nanocrystalline and polycrystalline ZrO_2, CeO_2 and Mn_3O_4 without and with 5 wt% Pd have been tested for methane oxidation. The nanocrystalline catalysts perform better than the polycrystalline catalysts on a weight basis. Nanocrystalline (n)-CeO_2 performs much better than n-ZrO_2, and both perform much better than polycrystalline (p)-CeO_2 and p-ZrO_2. 5 wt% Pd/n-ZrO_2 is the best catalyst tested, but 5 wt% Pd/n-CeO_2 performs nearly as well. A 20 wt% Ag/p-ZrO_2 and a 20 wt% Ag/n-ZrO_2 catalyst were also tested. The Ag yields a significant improvement over the bare supports, but not as much so as Pd.

1. INTRODUCTION

Methane is a powerful greenhouse gas and must be removed from the exhaust streams of natural-gas engines, but it is also the most difficult paraffinic hydrocarbon to oxidize catalytically due to the absence of C-C bonds. In vehicles which operate using natural gas, this removal can be accomplished using a catalytic converter equipped with the proper catalyst material. The use of Pd/Al_2O_3 catalysts in CH_4 oxidation has received the most attention, but they require high operating temperatures and often exhibit an unacceptable decay rate. Some studies have tested Pd-containing catalysts on supports other than Al_2O_3 (1-4). The results of these studies indicate that it may be possible to develop improved methane oxidation catalysts through the use of other supports such as ZrO_2 (3,4) CeO_2, Mn_3O_4, SnO_2 and Co_3O_4 (4). Based on these findings, Pd supported on polycrystalline (p-) ZrO_2, p-Mn_3O_4 and p-CeO_2 have been tested for CH_4 oxidation. Furthermore, due to the recent advancements in the preparation of nanocrystalline materials, their use as catalyst supports is also of interest and could lead to higher activities in catalytic processes due to their unique chemical and electronic properties. Therefore, Pd was also supported on nanocrystalline (n-) ZrO_2, n-Mn_3O_4 and n-CeO_2 supports and each was tested for CH_4 oxidation also. Ag supported on p- and n-ZrO_2 were also examined in order to determine if Ag could be used rather than the more costly Pd.

2. EXPERIMENTAL

Catalyst Supports

The nanostructured oxides used in this study were synthesized using the gas condensation method. In this process metals, suboxides or oxides are evaporated from a Joule-heated refractory crucible in a He atmosphere of 5-20 mbar. Nanoparticles, which are formed by homogeneous nucleation in the supersaturated vapor phase close to the crucible, are transported by a convective gas flow toward a LN_2-cooled stainless-steel plate where they are collected by thermophoretic forces. To prepare the nanocrystalline Mn_3O_4, ZrO_2 and CeO_2; Mn metal, ZrO and CeO_2 were evaporated respectively. The powders were then oxidized inside the synthesis chamber by replacing He with O_2. After scraping the nanopowders off the plate, they were more fully oxidized in flowing O_2 inside a tubular reactor heated to 300°C. The polycrystalline supports (99% minimum purity) were purchased from Aldrich Chemical Company or Alfa AESAR.

Catalyst Preparation

The Pd-containing catalysts were made using a Pd nitrate precursor purchased from Alfa AESAR. The oxide powders and $Pd(NO_3)_2$ were stirred into distilled water and heated to boil off the excess water. The remaining Pd-containing slurries were then calcined at 500°C for 4 hours. In order to test the catalytic properties of the bare support, the nanocrystalline and polycrystalline oxides were given the same treatments as the Pd-containing catalysts, but the metal precursors were not added. The Ag catalysts supported on p- and n-ZrO_2 were made in the same manner from an Ag_2O powder precursor purchased from Alfa AESAR.

Catalytic Activity Studies

The catalysts were tested for activity toward methane oxidation in a packed-bed reactor consisting of a glass tube using a reactant feedstream consisting of 10 at% O_2, 1 at% CH_4 in N_2 at a total pressure of 15 psia. All gases were of ultra-high purity. MKS mass-flow controllers were used to control the flow rate of each gas before mixing to form the feedstream. The reactant feed stream was passed through a 100 mg-catalyst bed at a flow rate of 33 standard cubic centimeters per minute (sccm). The reactor was enclosed in a Thermolyne Type 21100 tube furnace, and the reactor temperature was increased in increments beginning at room temperature. The compositions of the reactant and product feedstreams were tested using a Hewlett Packard 5790A Series gas chromatograph. A Tenax column containing Hayesep D packing material was purchased from Alltech to separate the components. H_2 and CO could be detected using this column but were not observed during any of the experiments indicating that partial oxidation of the methane did not occur.

3. RESULTS AND DISCUSSION

Methane conversion versus temperature data is shown in figure 1 for both nanocrystalline and polycrystalline ZrO_2. The bare ZrO_2 oxides are not very active for methane oxidation. Both the nanocrystalline and polycrystalline oxides exhibit similar light-off temperatures (the temperature at which the rate of activity increases indicating that the reaction is now diffusion limited) below 300°C, but n-ZrO_2 becomes considerably more active than p-ZrO_2 above 350°C. The n-ZrO_2 oxidizes about 40% of the CH_4 at 500°C whereas the p-ZrO_2 only oxidizes about 7% for the reaction conditions selected for this study. The greater activity of the n-ZrO_2 is most likely due to its greater surface area and therefore greater capacity for adsorption of the reactant gases. Methane conversion versus temperature data obtained from n-CeO_2 and p-CeO_2 are also shown in figure 1. These supports yield quite different results compared to the ZrO_2 catalysts. The p-CeO_2 material is a very poor catalyst for this reaction, but n-CeO_2 performs very well with a light-off temperature below 200°C, a rapid rise in activity with increasing temperature between 250 and 400°C and a slower rise from 400 to 500°C at which temperature complete conversion is achieved for the conditions used in this study. The difference in behavior between nanocrystalline versus polycrystalline is greater for CeO_2 than ZrO_2.

Similar studies were carried out using supported Pd because Pd has been most widely used for methane oxidation as discussed above. The Pd-containing catalysts are all 5 wt% Pd and calcined at 500°C. Neither the Pd loading nor the calcining temperature have been optimized in these studies. Methane conversion data as a function of temperature for Pd supported on nanocrystalline and polycrystalline Mn_3O_4 and CeO_2 are shown in figure 2. All four catalysts perform quite well with the CeO_2-supported catalysts performing better than the Mn_3O_4-supported catalysts. Pd/p-Mn_3O_4 has a light-off temperature below 200°C and exhibits complete conversion at about 450°C for the conditions used in this study. Pd/p-Mn_3O_4 and Pd/n-Mn_3O_4 perform similarly with Pd/n-Mn_3O_4 performing a little better than Pd/p-Mn_3O_4 below about 320°C and Pd/p-Mn_3O_4 performing better above this temperature. Both Pd/n-CeO_2 and Pd/p-CeO_2 exhibit complete conversion at about 360°C under the conditions used in this study. Pd/n-CeO_2 exhibits a low light-off temperature of 100°C whereas the light-off temperature for Pd/p-CeO_2 is above 200°C. Throughout the whole temperature range, Pd/n-CeO_2 performs better than Pd/p-CeO_2, but the rate of conversion slows more rapidly for Pd/n-CeO_2 at conversions above 93%.

Methane conversion versus temperature data are shown in figure 3 for n-ZrO_2, p-ZrO_2, 5 wt% Pd/n-ZrO_2, 5 wt% Pd/p-ZrO_2, 20 wt% Ag/n-ZrO_2 and 20 wt% Ag/p-ZrO_2. The influence of calcining temperature on n-ZrO_2 and p-ZrO_2 can be seen by comparing these data with the corresponding data shown in figure 1. For both the n-ZrO_2 and p-ZrO_2, the lower calcining temperature of 280°C (figure 3) results in significantly improved catalytic performance, and the n-

ZrO_2 is better in both cases. The addition of Pd results in excellent catalysts for both the n-ZrO_2 and p-ZrO_2 supports with complete conversions at about 350°C

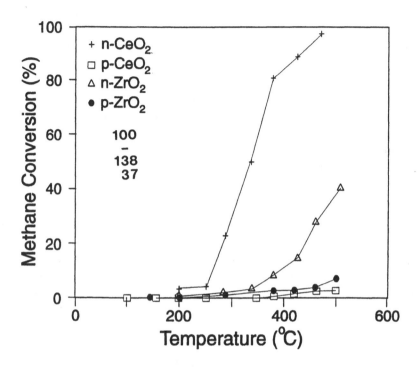

Figure 1   Methane conversion versus temperature for p-ZrO_2, n-ZrO_2, n-CeO_2 and p-CeO_2. The BET surface areas (m^2/g) are listed under the catalysts in the same order that the catalysts are listed in each figure.

for the conditions examined. Over the whole temperature range, the Pd/n-ZrO_2 catalyst performs better than the Pd/p-ZrO_2 catalyst. For example, at 250°C the Pd/n-ZrO_2 catalyst has an activity which is about twice that of the Pd/p-ZrO_2 catalyst, and at 300°C the conversion of Pd/n-ZrO_2 is about 96% while that of Pd/p-ZrO_2 is only about 68%. ZrO_2-supported Ag was also tested as a possible candidate for CH_4 oxidation. Ag is much less expensive than Pd and would therefore be an ideal substitute if active. The amount of CH_4 converted as a function of temperature using Ag supported on the p-ZrO_2 and n-ZrO_2 powders is shown in figure 3. The Ag/p-ZrO_2 catalyst oxidizes about 3% of the feed stream methane at 247°C and 13.7% at 292°C. The addition of Ag to the p-ZrO_2, under these reaction and preparative conditions, does indeed result in improved activity. Ag/n-ZrO$_2$ displays similar conversions to Ag/p-ZrO$_2$ at the lower temperatures, but above 292°C the n-ZrO$_2$ supported catalyst is superior to the Ag/p-ZrO$_2$ catalyst, oxidizing 15%

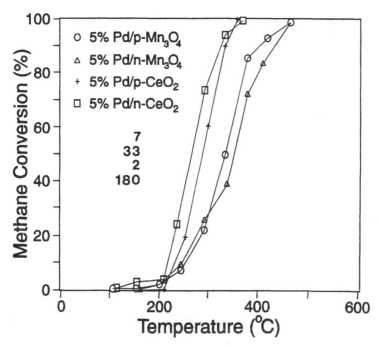

Figure 2   Methane conversion versus temperature for 5% Pd/p-Mn$_3$O$_4$, 5% Pd/n-Mn$_3$O$_4$, 5% Pd/p-CeO$_2$ and 5% Pd/n-CeO$_2$.

more CH$_4$ at 335°C. Complete conversion occurs at 468°C using the Ag/n-ZrO$_2$ sample while the Ag/p-ZrO$_2$ catalyst results in complete conversion above 509°C. Unfortunately, neither of these catalysts performs as well as the Pd-containing samples tested in this study, but the Ag-containing catalysts may be preferred in certain applications based on economic considerations. The BET surface areas are listed in the figures for most of the catalysts. The nanocrystalline catalysts have larger surface areas than the polycrystalline catalysts. On a weight basis the nanocrystalline catalysts perform better, but on a surface-area basis the polycrystalline catalysts perform better.

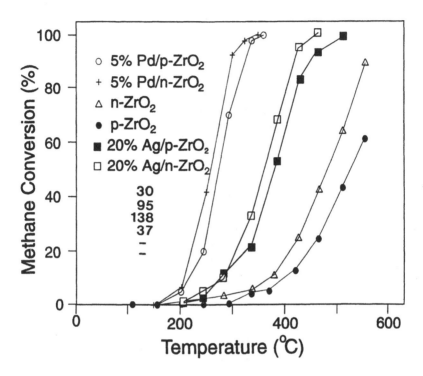

Figure 3   Methane conversion versus temperature for 5% Pd/p-ZrO$_2$, 5% Pd/n-ZrO$_2$, p-ZrO$_2$, n-ZrO$_2$, 20% Ag/p-ZrO$_2$ and 20% Ag/n-ZrO$_2$.

ACKNOWLEDGMENT

Support for this research was provided by the Petroleum Research Fund administered by the American Chemical Society through Grant PRF #29744-ACS, the National Science Foundation through NSF 9603409 INT, the German National Science Foundation (DFG) and the NSF/DAAD German-US exchange program.

5. REFERENCES

1.   C.F. Cullis and B.M. Willat, J. Catal. 86(1984)187.
2.   S. Seimanides and M. Stoukides, J. Catal. 98(1986)540.
3.   T.H. Ribeiro, M. Chow and R.A. Betta, J. Catal. 146(1994)537.
4.   W.S. Epling and G.B. Hoflund, to be published.

PROPERTIES OF PURE AND SULFIDED NiMoO₄ AND CoMoO₄ CATALYSTS: TPR, XANES AND TIME-RESOLVED XRD STUDIES

S. CHATURVEDI*, J.A. RODRIGUEZ*(1), J.C. HANSON*, A. ALBORNOZ**, AND J.L. BRITO**(1)

* Chemistry Department, Brookhaven National Laboratory, Upton, NY 11973, USA
** Centro de Quimica, IVIC, Caracas 1020-A, Venezuela

ABSTRACT

X-ray absorption near-edge spectroscopy (XANES) was used to characterize the structural and electronic properties of a series of cobalt- and nickel-molybdate catalysts ($AMoO_4.nH_2O$, α-$AMoO_4$, β-$AMoO_4$; A= Co or Ni). The results of XANES indicate that the Co and Ni atoms are in octahedral sites in all these compounds, while the coordination of Mo varies from octahedral in the α-phases to tetrahedral in the β-phases and hydrate. Time-resolved x-ray diffraction shows a direct transformation of the hydrates into the β-$AMoO_4$ compounds (following a kinetics of first order) at temperatures between 200 and 350 °C. This is facilitated by the similarities that the $AMoO_4.nH_2O$ and β-$AMoO_4$ compounds have in their structural and electronic properties. The molybdates react with H_2 at temperatures between 400 and 600 °C, forming gaseous water and oxides in which the oxidation state of Co and Ni remains +2 while that of Mo is reduced to +5 or +4. After exposing α-$NiMoO_4$ and β-$NiMoO_4$ to H_2S, both metals get sulfided and a $NiMoS_x$ phase is formed. For the β phase of $NiMoO_4$ the sulfidation of Mo is more extensive than for the α phase, making the former a better precursor for catalysts of hydrodesulfurization reactions.

INTRODUCTION

In this article, we describe recent studies that explore possible correlations among the structural, electronic, and catalytic properties of cobalt and nickel molybdates. Under atmospheric pressure, these compounds can exist as hydrates ($AMoO_4.nH_2O$, A= Co or Ni) or form the so-called α-$AMoO_4$ and β-$AMoO_4$ isomorphs. The catalytic properties of the cobalt and nickel molybdates are closely related to their structure [1-3]. The β-phase on $NiMoO_4$ is almost twice as selective for the dehydrogenation of propane to propene than the α-phase [1]. In a series of studies [2,3], the sulfided $NiMoO_4.nH_2O$ and $CoMoO_4.nH_2O$ compounds were found to be much better catalysts for the hydrodesulfurization (HDS) of thiophene than the corresponding sulfided α and β-isomorphs. The HDS activity of these systems increased in the following order: $\alpha < \beta <$ hydrate [2,3].

Synchrotron-based x-ray absorption near-edge spectroscopy (XANES) has emerged as a powerful tool for characterizing the electronic and structural properties of catalytic materials [4-6]. The $L_{II,III}$ edges of Co, Ni and Mo can provide information about the local site symmetry and oxidation state of these metals in oxides and sulfides [4,5,7]. Investigations at Brookhaven National Laboratory have recently established the feasibility of conducting sub-minute, time-resolved x-ray diffration (XRD) experiments under a wide variety of temperature and pressure conditions (-190 °C < T < 900 °C; P ≤ 45 atm) [8-10]. This important advance results from combining the high intensity of synchrotron

----

(1) Authors to whom correspondence should be addressed

radiation with rapid new parallel data-collection devices [8-10]. In this work, we use XANES, time-resolved XRD and temperature programmed reduction (TPR) to study the properties of pure and sulfided $CoMoO_4$ and $NiMoO_4$ catalysts.

EXPERIMENTAL

A series of $AMoO_4.nH_2O$ and $AMoO_4$ samples was prepared following the methodology described in refs. [2,3]. Samples of α-$NiMoO_4$ and β-$NiMoO_4$ were partially sulfided by exposing them to H_2S at 400 °C in a "U-shaped" pyrex reactor. The XANES spectra of the pure and sulfided catalysts were collected at the National Synchrotron Light Source on beamlines U7A (Co and Ni L-edges, Mo M-edge, O K-edge), X6A (Ni K-edge) and X19A (Mo L-edges, S K-edge). All these spectra were acquired at room temperature. The measurements at X6A and X19A were performed in the "fluorescence-yield mode", whereas the data at U7A were acquired in the "electron-yield mode".

The time-resolved powder diffraction patterns were collected on beamline X7B of the NSLS. The sample was kept in a quarz capillary and heated using a small resistance heater placed under the capillary [8-10]. A chromel-alumel thermocouple was used to measure the temperature of the sample. The x-ray powder diffraction patterns were accumulated on a flat image plate (IP) detector. In this experimental set-up, a continuos set of powder patterns can be obtained as a function of time or temperature [8-10]. The images collected on the IP were retrieved using a Fuji BAS2000 scanner.

The H_2O-TPD and H_2-TPR experiments were performed in a RXM-100 instrument from Advanced Scientific Designs Inc. Usually, the temperature of the sample was raised from 40 to 700 °C using a constant heating rate: 10, 20 or 30 °C/min. The reduction of the samples was carried out in a gas flow that consisted of 15% H_2+85% N_2 . The total flow rate was 50 ml/min. Chemical analysis of the desorbing products was performed using a quadrupole mass spectrometer (UTI 100C).

RESULTS

Dehydration of $CoMoO_4.nH_2O$ and $NiMoO_4.nH_2O$. The thermal stability of the $AMoO_4.nH_2O$ compounds was examined using TPD and time-resolved XRD. A detailed report of these studies will be presented elsewhere [11]. Upon heating of $AMoO_4.nH_2O$, evolution of gaseous water was seen in two different temperature ranges: 100-200 °C for reversibly bound H_2O; 200-400 °C for H_2O from the crystal structure. Figure 1 shows diffration patterns acquired during the heating of $NiMoO_4.nH_2O$ from room temperature to 600 °C. No significant changes in the diffraction patterns were seen at temperatures below 260 °C or above 360 °C. Around 300 °C the water that is bonded to the crystal structure desorbs, and one sees the collapse of the diffraction lines for the hydrate and the appearance of the lines for β-$NiMoO_4$. A similar transformation was observed for $CoMoO_4$ .nH_2O at temperatures between 200 and 250 °C. In general, the results of time-resolved XRD showed a direct change of the $AMoO_4.nH_2O$ hydrates into the β-$AMoO_4$ isomorphs without any intermediate α phase. This is probably facilitated by the similarities that the metal sites of the hydrates and β isomorphs have in their structural and electronic properties [11]. Using time-resolved XRD one can examine the kinetics of the hydrate→β transformation under isothermal conditions (i.e. for a given temperature monitor the changes in the intensity of the diffraction lines as a function of time) [8-10]. This type of studies reveal a kinetics of first-order in the hydrate with an activation energy that varies from 28 ($CoMoO_4.nH_2O$) to 32 kcal/mol ($NiMoO_4.nH_2O$).

Phase Transitions in $AMoO_4$. Experiments of XRD show that the α-$AMoO_4$→β-$AMoO_4$ transitions

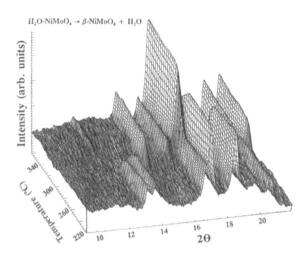

$H_2O\text{-}NiMoO_4 \rightarrow \beta\text{-}NiMoO_4 + H_2O$

Intensity (arb. units)

Temperature (°C)

2θ

Fig 1. Time-resolved x-ray diffraction patterns for the dehydration of $NiMoO_4 \cdot nH_2O$. The XRD experiments were carried out using a wavelength of 0.957 Å and a heating rate of 4 °C/min.

occur at much higher temperatures than the hydrate→β-$AMoO_4$ transformations ($\Delta T \approx 150$ °C in $CoMoO_4$ and 280 °C in $NiMoO_4$). The activation energy for the α-$NiMoO_4$→β-$NiMoO_4$ transition is ~ 40 kcal/mol larger than that for the $NiMoO_4 \cdot nH_2O$→β-$NiMoO_4 + nH_2O$ reaction. The larger activation energy probably reflects the change in the coordination of Mo (O_h→T_d) that occurs during the α→β transition. In addition, the results of XANES show a substantial change (~ 1.5 eV) in the splitting of the Mo 4d orbitals during this transition.

Reaction of Hydrogen with $AMoO_4$ and $AMoO_4 \cdot nH_2O$. Figure 2 shows spectra for the reduction of α-$NiMoO_4$, α-$CoMoO_4$, and β-$CoMoO_4$ in a gas mixture of H_2 and N_2. For α-$NiMoO_4$, the uptake of hydrogen begins around 400 °C and two peaks for desorption of water are seen at ~ 550 and 650 °C. For the cobalt molybdates, the reaction with hydrogen starts around 500 °C and the first water desorption peak appears at 610-640 °C. The data for the consumption of hydrogen agree well with results of previous H_2-TPR studies [3,12]. In theTPR of $NiMoO_4 \cdot nH_2O$ (not shown), the first peak for the consumption of H_2 appears at ~ 400 °C, almost 150 °C below the corresponding peak of α-$NiMoO_4$ [3,12]. The TPR spectra for $CoMoO_4 \cdot nH_2O$ were very similar to those displayed in Fig 2 for β-$CoMoO_4$. In the water-desorption spectrum of $CoMoO_4 \cdot nH_2O$, there were two "extra" peaks between 100 and 350 °C as a result of the $CoMoO_4 \cdot nH_2O$ → β-$CoMoO_4 + nH_2O$ transformation.

In a set of experiments the $AMoO_4$ and $AMoO_4 \cdot nH_2O$ compounds were partially reduced and their new properties examined using XRD and XANES. The samples were put in the reaction cell under a 15%-H_2+85%-N_2 mixture (flow rate= 50 cc/min) and the temperature was ramped from 15 to 600 °C. After that, the samples were cooled to room temperature and exposed to oxygen from the atmosphere. In subsequent TPR experiments, we found that the "reduced" cobalt molybdates did not

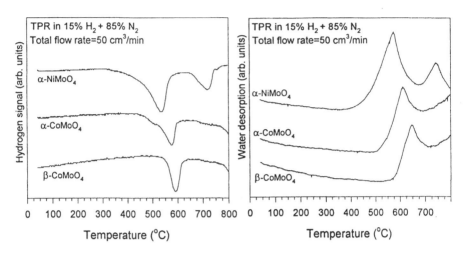

Fig 2 Hydrogen consumption (left) and water desorption (right) during the H_2-TPR of α-NiMoO$_4$, α-CoMoO$_4$ and β-CoMoO$_4$. Heating rate= 20 °C/min

react with hydrogen at temperatures between 25 and 600 °C. In the case of the "reduced" nickel molybdates, the TPR showed the consumption of a small amount of H_2 from 400 to 500 °C.

The "reduced" β-CoMoO$_4$ and CoMoO$_4$.nH$_2$O samples had identical x-ray powder patterns at room temperature. These diffraction patterns were well defined and can be assigned to a mixture of CoMoO$_3$ and another oxide. We examined the thermal stability of these systems. The results of time-resolved XRD (see Fig 3) indicate that around 600 °C they undergo a change in their structural

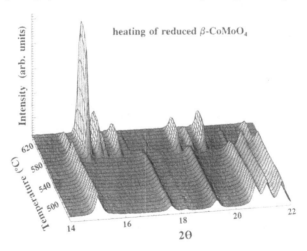

Fig 3 Time-resolved x-ray diffraction patterns obtained during the heating of "reduced" β-CoMoO$_4$. Heating rate= 2 °C/min; λ= 0.92 Å.

geometry and the typical diffraction pattern for β-CoMoO$_4$ appears. At this point, however, the "reduced" samples are not the typical β-CoMoO$_4$ isomorph. They exhibited a black color, and were very rich in oxygen vacancies showing a conductivity larger than that of β-CoMoO$_4$.

The results of XANES for the L$_{II,III}$ edges of Co and Ni show no change in the oxidation state of these metals after the partial reduction of the cobalt and nickel molybdates. The corresponding O K-edge spectra indicate that there is a change in the average oxidation state of Mo (6+ → 5+) that has a direct impact on the O(1s)→Mo(4d) electronic transitions. In a similar way, after examining the peak positions in the Mo M$_{III}$ edges of the "reduced" molybdates (see Fig 4), one finds an average oxidation state of +5 for Mo.

Sulfidation of Nickel Molybdates. After exposing α-NiMoO$_4$ and β-NiMoO$_4$ to H$_2$S at 400 °C, the S K-edge spectra showed the presence of two types of sulfur species in the samples. One was associated with the metal sites (NiS$_y$ and MoS$_x$) and had a formal oxidation state of -2. The other had a formal oxidation state of +6 and was associated with the oxygen sites (SO$_4^{2-}$ species). An analysis of the extended structure in the Ni K-edge spectrum suggests that a NiMoS$_x$ phase was present in the sulfided molybdates. The main peak positions in the Mo L$_{II}$- and Mo$_{III}$-edge spectra indicate that in the β phase of NiMoO$_4$ the sulfidation of Mo was more extensive than in the α phase, with β-NiMoO$_4$+S showing a peak position that is closer to that of MoS$_2$ (see Fig 4). This difference in reactivity makes β-NiMoO$_4$ a better precursor than α-NiMoO$_4$ for the preparation of hydrodesulfurization (HDS) catalysts [2,13].

DISCUSSION AND CONCLUSION

Our studies show that time-resolved XRD and XANES are powerful tools for examining the

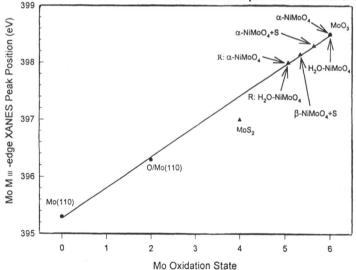

Fig 4 Mo oxidation state from Mo M$_{III}$-edge XANES spectra. "R" and "S" denote partially reduced (with H$_2$) or sulfided (with H$_2$S) samples.

structural, electronic and chemical properties of novel mixed-metal oxide catalysts. The results of time-resolved XRD show that the $AMoO_4 \cdot nH_2O$ compounds transform directly into the corresponding β-isomorphs without any intermediate phase. The relatively low temperatures at which this process occurs (240-340 °C) make the hydrates very good precursors for the preparation of β-$AMoO_4$ catalyst used in the partial oxidation of hydrocarbons.

The K- and L-edges of Co and Ni indicate that there are no major changes in the electronic properties of these metals when going from α-$AMoO_4$ to β-$AMoO_4$. On the other hand, the $L_{II,III}$-edges of Mo and the K-edge of O showed large variations in the electronic properties of Mo as the coordination of this metal varies from octahedral (α-$AMoO_4$) to tetrahedral (β-$AMoO_4$), or after partial reduction of the molybdates with H_2. These changes in the electronic properties of Mo correlate with changes in catalytic activity reported for the cobalt and nickel molybdates [1-3]. In the tetrahedral environment of the β-phase, a Mo^{6+} ion is coordinatively unsaturated and in overall its empty 4d orbitals are less destabilized than in an octahedral field. This could favor bonding interactions between the Mo 4d orbitals and the occupied orbitals of hydrocarbons (partial oxidation reactions) or the sulfur lone-pairs of thiols and sulfides (HDS reactions), making the β-phase more chemically active than the α-phase.

ACKNOWLEDGEMENTS

This work was supported by the US-DOE (DE-AC02-76CH00016). One of the authors (JLB) gratefully acknowledges a grant from ACS that made possible a visit to BNL.

REFERENCES

1. C. Mazzocchia, C. Aboumrad, C. Diagne, E. Tempesti, J.M. Herrmann and G. Thomas, Catal. Lett. **10**, p. 181 (1991).
2. J.L. Brito, A.L. Barbosa, A. Albornoz, F. Severino and J. Laine, Catal. Lett. **26**, p. 329 (1994).
3. J.L. Brito and A.L. Barbosa, J. Catal., **171**, p. 467 (1997).
4. Proceedings of the 8th International Coference on X-ray Absorption Fine Structure (XAFS VIII), Physica B, **208&209** (1995).
5. J.G. Chen, Surf. Sci. Reports, **30**, p. 1 (1997).
6. J.M. Thomas and W.J. Thomas, The Principles and Practices of Heterogeneous Catalysis (VHC: New York, 1997).
7. S.R. Bare, G.E. Mitchell, J.J. Maj, G.E. Vrieland and J.L. Gland, J. Phys. Chem. **97**, p. 6048 (1993).
8. P. Norby, A.N. Christensen, and J.C. Hanson, Studies in Surf. Sci. Catal. **84**, p. 179 (1994).
9. A. Nordlund-Christensen, P. Norby and J.C. Hanson, J. Solid State Chem. **114**, p. 556 (1995).
10. P. Norby and J.C. Hanson, Catal. Today, submitted.
11. J.A. Rodriguez, S. Chaturvedi, J.C. Hanson, A. Albornoz and J.L. Brito, to be published.
12. J.L. Brito, J. Laine and K.C. Pratt, J. Mater. Sci. **24**, p. 425 (1989).
13. S. Chaturvedi, J.A. Rodriguez and J.L. Brito, to be published.

RELATIONSHIP BETWEEN CATALYTIC ACTIVITY AND ELECTRICAL CONDUCTIVITY IN THE NONSTOICHIOMETRIC NICKEL MANGANITE SPINELS

Christel Laberty[1], Pierre Alphonse [2] and Abel Rousset [2]
[1] Department of Chemical Engineering and Materials Science, University Of California at Davis, Davis CA 95616, USA,
[2] Laboratoire de Chimie des Matériaux Inorganiques ESA5070, 118 Route de Narbonne, 31062 Toulouse Cedex, France.

Abstract

Nonstoichiometric nickel manganite spinels $Ni_xMn_{3-x}\square_{3\delta/4}O_{4+\delta}$, have been synthesised by calcination of mixed oxalates in air at 350°C. The variation of the electrical conductivity σ with partial pressure of O_2 shows that these oxides are n-type semiconductors; σ also varies with the nickel content and has a maximum at x = 0.6. The intrinsic catalytic activity of these oxides for CO/CO_2 conversion varies with nickel content and the most active catalyst is at x=0.6 where the conversion starts at room temperature. The variation of the catalytic activity and the electrical conductivity and the nickel amount are correlated. Apparent activation energies are very low (less than 20 kJ/mol) and remain the same for all these mixed oxides. Similarly, the reaction order with respect to O_2 and CO does not depend on the nickel content (order/$O_2 \approx 0$; order/CO ≈ 0.6). A reaction mechanism involving formation of oxygen ions adsorbed and carbonate species is discussed.

Introduction: Nickel manganite spinels, widely studied for their application as thermistor ceramics, are usually prepared by thermal decomposition of mixed manganese-nickel oxalates in air at 900°C. These oxides have a low specific surface area (1 m^2 g^{-1}) and are stoichiometric (1,2). Recently, thermal decomposition of manganese-nickel oxalate at low temperature (T < 350°C) has been investigated (3-5) and the formation of cation deficient spinels $Ni_xMn_{3-x}\square_{3\delta/4}O_{4+\delta}$ has been shown. These oxides are finely divided materials. Their BET specific surface area is close to 100 m^2g^{-1} and the average diameter of their crystallites is about 10 nm (6). The cation deficiency δ has been evaluated by TGA under controlled atmosphere. It varies with nickel content and reaches a maximum between x = 0.4 and x = 0.6 (6,7). The marked nonstoichiometry of these oxides can be linked to their high reactivity towards molecules like O_2, CO or hydrocarbons. The goal of this study is to correlate the catalytic activity of CO oxidation and electrical conductivity at different temperatures and under different atmospheres with the nickel content.

Synthesis: Nonstoichiometric nickel manganites were prepared by decomposition of mixed oxalates at 350°C in air for 6 hours. The oxalates $Ni_{x/3}Mn_{(3-x)/3}(C_2O_4).2H_2O$ ($0 \leq x \leq 1$) were obtained at room temperature by coprecipitation of aqueous solution of nickel and manganese nitrates with a solution of ammonium oxalate. After one hour, the mixture was filtered, washed several times with deionized water and dried at 25°C in air.

Structure: X-ray powder diffraction patterns show that the cation deficient nickel manganites are poorly crystallised, but the most intense lines of the spinel structure clearly appear. Significant information on the structural architecture of these materials have been obtained from a Wide Angle X-ray Scattering (WAXS) study (8), which show that in the amorphous state, the structural arrangement of manganese, nickel and oxygen atoms follows a spinel structure. Moreover, this study points out that oxalate decomposition leads directly to a spinel structure with a very high content of vacancies essentially localized in tetrahedral sites.

Nonstoichiometry determination: To evaluate the nonstoichiometric level δ and to determine the oxidation state and the content of the different cations as a function of nickel content, thermogravimetric analyses under controlled atmospheres (Ar, H_2, air) were done. Heated in Ar, the nonstoichiometric compounds lose oxygen in several steps leading to the stoichiometric oxide: the first step is assigned to reduction of Ni^{3+} to Ni^{2+}, while the others correspond to the reduction of a part of the Mn^{4+} and Mn^{3+} cations (9). Temperature programmed reduction (TPR) of these oxides in H_2 leads in several steps to the formation of a mixture of nonstoichiometric $MnO_{1-\delta}$ and metallic nickel. Similar to the reaction in Ar, the first step is attributed to the reduction of Ni^{3+} (7). The experimental weight loss evaluated from TGA experiments in H_2 and Ar and TGA studies of mixed oxalate decomposition (6), allow the evaluation of the vacancy content (δ). The values are plotted as a function of the nickel content in figure 1. The vacancy levels in these oxides are high and vary with the nickel amount showing a maximum between x=0.4 and x=0.6.

Fig 1. Variation of δ with x in $Ni_xMn_{3-x}\square_{3\delta/4}O_{4+\delta}$

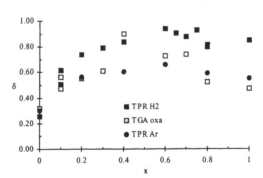

Cation distribution: Because of the strong preference of the Ni^{2+} cation for octahedral sites, stoichiometric nickel manganites are inverse spinels with the cation arrangement: $Mn^{2+}\left[Ni^{2+}Mn^{4+}_xMn^{3+}_{2-2x}\right]O^{2-}_4$. Ni^{2+} on the octahedral sites causes the existence of $Mn^{3+/4+}$ redox couples on the same lattice sites. These couples are the reason for the good electrical conductivity of nickel manganites. Usually, normal manganite spinels are all insulators. For x=1, the cation distribution would be: $Mn^{2+}\left[Ni^{2+}Mn^{4+}\right]O^{2-}_4$. This arrangement does not account for the good conductivity of this oxide which implies that, for high nickel content (x > 0.6), a part of the Ni^{2+} cations must be in tetrahedral sites giving the distribution:

$Ni^{2+}_{(1-2\lambda)x}Mn^{2+}_{2\lambda x}[Ni^{2+}_{2\lambda x}Mn^{4+}_{2\lambda x}Mn^{3+}_{2-4\lambda x}]O^{2-}_4$ where $0.25 \leq \lambda \leq 0.5$ depending on total nickel content and thermal history of sample (10).

Cation deficient nickel manganites contain a high number of vacancies and a WAXS study (8) has shown that these vacancies are mainly localized in tetrahedral sites. Moreover, interpretation of TGA profiles in H_2 lead to suppose the existence of Ni^{3+} in these oxides. Among Ni^{3+}, Ni^{2+}, Mn^{3+} and Mn^{4+}, Ni^{2+} has the lowest octahedral preference (11). A combination of this statement with the value of δ and the amount of Ni^{3+} evaluated from TGA analyses allows to determine the cation distribution in nonstoichiometric nickel manganites. For x=0.2, the cation distribution would be:

$$Mn^{2+}_{0.61}\square_{0.39}[Ni^{3+}_{0.15}Ni^{2+}_{0.02}Mn^{3+}_{1.02}Mn^{4+}_{0.80}]O^{2-}_4$$

Electrical conductivity: To measure the electrical conductivity σ, the sample powder was hold between two Ni grids and pressed by a spring. The resistivity was recorded between 25-100ºC under N_2 and O_2. The maximum of the curve $Ln(\sigma) = f(x)$ is observed at x = 0.6 (fig 3). Similar behavior has already been noticed with stoichiometric oxides (9). Assuming that the conduction mechanism in nonstoichiometric nickel is identical to the one in stoichiometric oxides (9), it is not surprising that the maximum of the conductivity corresponds to x = 0.6 since this oxide has the number of couple maximum.

The variation of σ with the partial pressure of oxygen at 50 ºC and 100 ºC is reported in Table 1. For nonstoichiometric oxides, σ slightly decreases with increasing O_2 concentration up to a high partial pressure of O_2, then σ becomes independent of the O_2 concentration. The decrease of σ with increasing O_2 concentration shows that nonstoichiometric manganites are n-type semiconducting. The surface adsorption of O_2 depletes the surface layer of electrons and can be described by the following equation: $O_2 + e^- \rightarrow \underline{O_2^-}$

On the contrary, for manganese oxide the electrical conductivity (σ) increases with increasing O_2 concentration characteristic of p-type semiconductors. The O_2 adsorption increases the number of charge that is in this case positive holes: $O_2 \rightarrow \underline{O_2^-} + h^+$

For both types of oxides, the chemisorption of negatively charged oxygen species does result in the formation of an electrical double layer (12). This layer inhibits the extraction of electrons and the adsorption is stopped shown by the independence of the electrical conductivity (σ) with the O_2 concentration.

Table 1
Variation of σ with O_2 concentration for $Ni_xMn_{3-x}\square_{3\delta/4}O_{4+\delta}$

x	T(°C)	O_2 concentration (%)						
		0	1	3	7	10	20	30
0	50	$2.51.10^{-8}$	$2.53.10^{-8}$	$2.62.10^{-8}$	$2.71.10^{-8}$	$2.76.10^{-8}$	$2.79.10^{-8}$	$2.79.10^{-8}$
	100	$3.02.10^{-7}$	$3.09.10^{-7}$	$3.14.10^{-7}$	$3.27.10^{-7}$	$3.47.10^{-7}$	$2.49.10^{-8}$	$3.50.10^{-8}$
0.6	50	0.54	0.52	0.52	0.51	0.50	0.50	0.50
	100	0.36	0.34	0.34	0.32	0.30	0.30	0.30
1	50	$1.48.10^{-4}$	$1.47.10^{-4}$	$1.46.10^{-4}$	$1.44.10^{-4}$	$1.40.10^{-4}$	$1.40.10^{-4}$	$1.40.10^{-4}$
	100	$2.40.10^{-2}$	$2.34.10^{-2}$	$2.32.10^{-2}$	$2.31.10^{-2}$	$2.29.10^{-2}$	$2.29.10^{-2}$	$2.29.10^{-2}$

Catalytic activity: Catalytic tests were made in a tubular fixed bed flow reactor at air pressure, in the temperature range 50-250°C, with a residence time V/F = 0.01 s and feeding reacting mixtures of $CO-O_2-N_2$ (1%-2%-97%). Reactants and products were analysed by GC using two columns: 13x molecular sieve for N_2, O_2 and CO and porapak Q for CO_2. Preliminary tests showed variation of CO/CO_2 conversion with specific surface area. Because of the difficulties to obtain all the catalyst samples with similar specific surface area, intrinsic activity α was used (expressed in $mole.s^{-1}.m^{-2}$).

For the most active oxides, the conversion starts at room temperature. Figure 2 shows the variation of the intrinsic activity at 120°C with nickel content. Two sets of results are presented, labelled N_2 or O_2 pre-treated, according to the atmosphere in which the catalyst was heated 1 hour at 300°C before catalytic tests. The activity of the mixed oxides is larger than the activity of the monometallic oxides and the most active mixed oxide corresponds to x=0.6. On the other hand, in all cases, O_2 pre-treated samples are the most active. The variation of α with x compares well to the variation of σ with x. A comparison between the variation of the catalytic activity and the nonstoichiometry level δ with x shows that the oxide the most active is the one containing greatest level of vacancy.

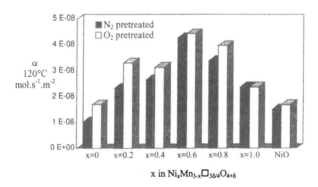

Fig.2. Variation of nonstoichiometric oxides catalytic activity with the nickel content

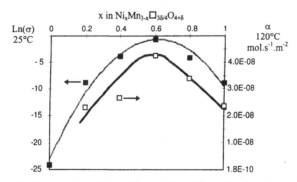

Fig. 3. Comparison between the variation of the catalytic activity and the electrical conductivity

Kinetics: The values of the apparent activation energy of different oxides, calculated from the variation of the catalytic activity with temperature (table 2), are very low and mass-transfer limitations could be suspected. But, in our experimental conditions, the intrinsic activity does not change with the flow rate showing no mass-transfer limitations. The calculated values of the apparent activation energy are independent on the nickel content and the pre-treatment, except for manganese oxide. In this case, the value observed is larger almost by a factor of 2 than that of the nonstoichiometric nickel manganite. These results suggest that nickel cations play a greater role in the reaction mechanism than Mn cations.

The reaction order with respect to O_2 of nonstoichiometric oxides are similar and near zero (table 3) otherwise the reaction orders with respect to CO are different and vary between 0.5 –0.7.

Table 2
Variation of activation energy and reaction order with nickel content (x)

x	0	0.2	0.4	0.6	0.8	1	NiO
$E(kJ.mole^{-1})$ O_2 pretreat	33	17	17	17	17	17	17
$E(kJ.mole^{-1})$ N, pretreat	29	17	17	17	17	17	17
order/CO	0.5	0.6	0.6	0.7	0.6	0.5	0.9
order/O_2	0.1	0.2	0.1	0.2	0.1	0.2	0.1
global order	0.7	0.8	0.9	0.7	0.7	0.8	1.0

According to these results, a reaction mechanism for the catalytic oxidation of CO can be discussed. The reaction orders suggest that both CO and O_2 are chemisorbed and the variation of electrical conductivity with the partial pressure of O_2 shows the adsorption of O_2 as O_2^- type (see equation 1). On the other hand, investigations by IR spectroscopy of the surface species adsorbed on these nonstoichiometric oxides (13) during the catalytic process have shown that near room temperature, reaction with flowing gas mixtures CO in He or CO+O_2 in He, leads to the formation of carbonate species and gaseous CO_2. Moreover, when the reactant gas mixture is replaced by pure He, most of the carbonate species remain on the surface although with a flowing mixture of CO in He or O_2 in He, the carbonate quickly disappears.

Formation of carbonate species probably arises from the interaction between gaseous CO and oxygen adsorbed species:

$$CO + \underline{O_2^-} \rightarrow \underline{CO_3^-} \qquad \text{fast} \qquad (2)$$

Since CO or O. promotes the decomposition of the surface carbonate structure,

$$CO + \underline{CO_3^-} \rightarrow 2\,\underline{CO_2} + e^- \qquad \text{slow} \qquad (3)$$
$$O_2 + 2\,\underline{CO_3^-} \rightarrow 2\,\underline{CO_2} + 2\,\underline{O_2^-} \qquad \text{slow} \qquad (4)$$

the last step is the CO_2 desorption. However, the rate of disappearance of CO_2 is strongly dependent on the temperature:

$$\underline{CO_2} \rightarrow CO_2 \qquad (5)$$

The value of activation energy seems to indicate that reactions (1) and (3) mainly occur on Ni cations and then could be written as:

$$Ni^{2+} + O_2 \rightarrow Ni^{3+} + \underline{O_2^-} \qquad \text{fast} \qquad (1)$$
$$Ni^{3+} + CO + \underline{CO_3^-} \rightarrow Ni^{2+} + 2\,\underline{CO_2} \qquad \text{slow} \qquad (3)$$

Conclusion: Nonstoichiometric nickel manganite spinels $Ni_xMn_{3-x}\square_{3\delta/4}O_{4+\delta}$, synthesised by calcination of mixed oxalates in air at 350°C, are n-type semiconductors. The electrical conductivity varies with the nickel content and has a maximum at x = 0.6 where the couple numbers (Ni^{2+}/Ni^{3+} and Mn^{3+}/Mn^{4+} in octahedral sites) are the highest. The intrinsic catalytic activity of these oxides, for CO/CO_2 conversion, varies also with nickel content and the most active catalyst observed for x=0.6 starts the conversion of CO at room temperature. A correlation is found between the variations of the catalytic activity and the electrical conductivity with nickel content. Apparent activation energies of these oxides are very low (less than 20 kJ/mol) and independent on the nickel content. Similarly the reaction order with respect to O_2 and CO appear little dependent on nickel amount (order/$O_2 \approx 0$; order/CO ≈ 0.6). A mechanism for oxidation of CO involving the formation of oxygen charged and carbonate species is proposed

References:

1. E. Jabry, G. Boissier, A. Rousset, R. Carnet, A. Lagrange, *J. Phys. Colloq., C1*, **47**, 843 (1986).

2. B. Gillot, J. L. Baudour, F.Bouree, R. Metz, R. Legros, A. Rousset, *Solid State Ionics,* **58**, 155 (1992).

3. X. X. Tang, A. Manthiram, J. B. Goodenough, *J. Less-Common Met.*, **156**, 357 (1989).

4. A. Feltz, J. Töpfer, *Z. Anorg. Allg. Chem.*, **576**, 71 (1989).

5. J. Töpfer, J. Jung, *Thermo. Acta.*, **202**, 281 (1992).

6. C. Laberty, P. Alphonse, J. J. Demai, C. Sarda, A. Rousset, *Mat. Res. Bull.*, **32**, 2, 249, (1996).

7. C. Laberty, J. Pielaszek, P. Alphonse, A. Rousset, *submitted to publication.*

8. C. Laberty, M. Verelst, P. Lecante, P. Alphonse, A. Mosset , A. Rousset, *J.Solid State Chem.,* **129**, 271, (1997).

9. C. Laberty, Thesis, Toulouse France, October 1997.

10. S.Fritsch, R. Legros, A. Rousset, Advances in Science and Technology, vol 2D. Ceramics : Charting the future, P. Vincenzini Ed, Techna Srl, 2565, (1995).

11. A. Navrotsky, O.J. Kleppa, J. Inorg. Nucl. Chem., **29**, 2701, (1967)

12. Th. Wolkenstein, Théorie électronique de la catalyse sur les semi-conducteurs, Masson Ed., Paris, (1961)

13. C. Laberty, C. Mirodatos, P. Alphonse, A. Rousset, to be published.

THE PARTIAL OXIDATION OF METHANOL BY MoO3(010) SURFACES WITH CONTROLLED DEFECT DISTRIBUTIONS

RICHARD L. SMITH and GREGORY S. ROHRER
Carnegie Mellon University,
Department of Materials Science and Engineering
Pittsburgh, Pennsylvania 15213-3890

ABSTRACT

Atomic force microscopy has been used to determine how $MoO_3(010)$ surfaces with controlled defect populations evolve during reactions with $MeOH/N_2$ mixtures. The structural evolution of the freshly cleaved surface is compared to surfaces reduced in $10\%H_2/N_2$ at 400 °C. While the freshly cleaved surfaces are flat and nearly ideal, the reduced surfaces contain voids bounded by step loops where Mo atoms in reduced coordination are found. In both cases, reacting the $MoO_3(010)$ surface with $MeOH/N_2$ mixtures between 300 and 400 °C leads to H intercalation and the nucleation of acicular precipitates of H_xMoO_3. On the freshly cleaved surface, the H-bronze phase precipitates uniformly. On the reduced surface, the precipitates form preferentially at the void edges, where undercoordinated Mo atoms are found.

INTRODUCTION

Molybdenum trioxide (MoO_3) is a common component of catalysts used for the partial oxidation of hydrocarbons. Multi-component materials of practical importance include the Bi-Mo-O catalyst used for propene oxidation and ammoxidation and the Fe-Mo-O catalyst used for the partial oxidation of methanol to formaldehyde. MoO_3 itself has been the focus of many experimental investigations, particularly because many reactions catalyzed by the oxide are thought to be surface-structure sensitive [1-4]. It has been proposed that unsaturated surface sites, where Mo atoms have fewer than six nearest neighbors, are preferential reaction sites for alcohol oxidation [3]. The objective of the experiments described in this paper was to test this idea by intentionally introducing a well-characterized population of undercoordinated Mo on the $MoO_3(010)$ surface and directly observing the morphological evolution during reactions with alcohols.

MoO_3 has the layered structure illustrated in Fig. 1. Each layer is composed of two corner-sharing octahedral nets that link by sharing edges along [001]. Adjacent layers along [010] are linked only by weak van der Waals forces and, when cleaved at the gap between these layers, two identical O-terminated surfaces are created. While the Mo atoms in the ideal surface layer retain their bulk coordination, Mo atoms in alternative coordinations exist at step edges. If a step is created along [001] by breaking Mo-O-Mo linkages, one half of the metal atoms are left in a five-coordinate site. If, on the other hand, a step is created along [100], two Mo-O-Mo linkages are

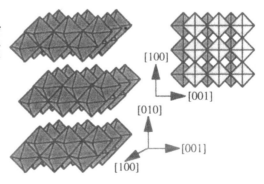

Figure 1. Polyhedral representation of the MoO_3 structure. MoO_3 has a primitive orthorhombic cell (Pbnm) and its lattice parameters are **a**=3.963 Å, **b**=13.856 Å, **c**=3.6966 Å.

[100]

[001]

[010]

[001]

[1̄00]

broken for each Mo atom and the average Mo coordination number along the step must be 5. In this case, the average might be made up of equal numbers of 4- and 6-coordinate Mo.

Experimentally, it is possible to produce undercoordinated sites with reducing treatments at 400 °C in H_2/N_2 mixtures [5]. Under these conditions, voids, bounded by step loops, form on the surface. We have used atomic force microscopy (AFM) to characterize both the freshly cleaved surface and surfaces containing voids. In this paper, we describe the evolution of these two characteristic surfaces during reactions in $MeOH/N_2$ mixtures. In each case, H_xMoO_3 precipitates are observed to form as a product of the reaction, but the distribution of the precipitates differs. The mechanisms by which the undercoordinated Mo at the edges of the voids might affect the transfer of protons from the alcohol to MoO_3 are discussed.

EXPERIMENTAL

Single crystals of MoO_3 were grown by chemical vapor transport in sealed, evacuated quartz ampoules [6]. The surface reactions were performed in a quartz tube with flowing gas at 1 atm. A magnetic transfer rod (consisting of a permanent magnet sealed within a quartz rod) was used to move the samples into and out of the hot zone of the furnace. The single crystals were mounted on a steel disc by spot welding a thin strip of Ta foil across the basal facet. The steel disc was attached to the transfer rod by chromel wire spring clips. After cleaving the crystal and sealing the reaction tube, the system was purged by alternately evacuating and backfilling with the reactant gas mixture three times. The samples were then transferred to the hot zone of the furnace and reacted for a pre-determined time at temperature.

To create surface voids with undercoordinated Mo atoms, as-received 10%H_2/N_2 (forming gas) with a nominal H_2O concentration of 20 ppm was used to reduce the samples at 400°C in a 350 cc/min flow of the gas. Samples were then transferred to the cool zone of the reaction tube and allowed to sit under a flow of forming gas for approximately 30 min while the hot zone was cooled to the desired temperature for the MeOH reaction. Subsequently, the gas composition was changed to a $MeOH/N_2$ mixture and the sample was transferred back to the hot zone for reaction under a 200 cc/min flow. The N_2 (Prepurified, Matheson) carrier gas was dried with $CaSO_4$ (Drierite) and a liquid N_2 trap prior to being saturated in a methanol (99.9+%, Aldrich) bubbler. Following reaction, the MoO_3 samples were cooled to room temperature under a flow of the reactant gas mixture and then transferred to an AFM housed in a glove box with an atmosphere of continuously-purified Ar (O_2 and $H_2O < 5$ ppm). In most cases, the samples were transferred directly to the glove box without exposing them to the ambient atmosphere. However, the microstructural changes observed on surfaces which were momentarily exposed to the ambient did not differ from those of samples which were not exposed to air.

X-ray diffraction (XRD) was used to identify the phase that formed on the $MoO_3(010)$ surfaces during the reactions with methanol. First, a number of crystals were reacted simultaneously in a silica boat. To selectively probe the near surface region, the reacted (010) surface layers (thickness < 20 μm) were cleaved from the crystals with adhesive tape, pulverized, and immediately subjected to XRD analysis. The small quantity of powder obtained with this procedure was mounted on a glass slide covered with a layer of double-sided tape. At least one of the crystals from each batch was characterized with AFM to ensure that the surface structure could be directly correlated with the X-ray data. In some cases, parallel experiments with powders were also performed. All X-ray data were recorded in the ambient on a Rigaku Θ-2Θ diffractometer with Cu K_α radiation. Lattice parameters were refined using a least squares method.

RESULTS

The freshly cleaved (010) surface is characterized by atomically flat terraces separated by steps whose heights are always an integer multiple of 7 Å (7 Å ≈ **b**/2). Steps with heights greater than 28 Å (2 unit cells) are rarely observed. The step edges are almost exclusively oriented along the [001] axis of the crystal, as can be seen in the characteristic AFM image in Fig. 2. Creation of steps of this orientation requires the minimum number of broken bonds per unit length of step.

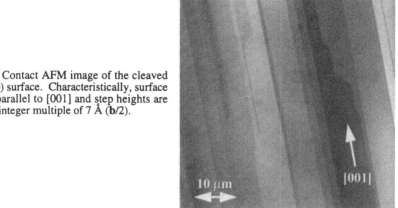

Figure 2. Contact AFM image of the cleaved MoO₃(010) surface. Characteristically, surface steps run parallel to [001] and step heights are always an integer multiple of 7 Å (**b**/2).

Reactions with MeOH/N₂ mixtures between 300 and 340 °C lead to easily observed changes in the surface structure. Within 10 min of reaction at 330 °C, the surface is decorated with a cross hatched pattern of acicular precipitates oriented along <203> (see Fig. 3a). The precipitates continue to grow with increasing reaction times and after extended reactions (t = 20 min), they coalesce and assume a second habit parallel to [001] (see Fig. 3b). These structures along [001] are very straight and can have lengths up to several millimeters. The precipitates appear white in AFM images because they rise out of the MoO₃ matrix.

Powder XRD experiments unambiguously identified the <203> precipitates. The powder XRD patterns from reacted surface layers with microstructures similar to the one shown in Fig. 3b clearly show that the surface is composed of a two phase mixture. In Fig. 4, a portion of a representative pattern is compared to that of unreacted MoO₃ so that the peaks from the second phase can be easily identified. The new peaks in the XRD pattern of the reacted material, which we associate with the <203> precipitates, can be indexed to a C-centered orthorhombic cell

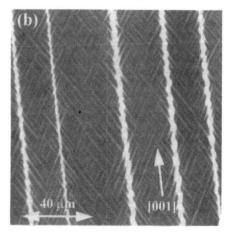

Figure 3. Contact AFM images of MoO₃(010) surfaces after reaction with N₂ saturated with MeOH (25 °C) for (a) 8 and (b) 20 min. at 330 °C. The black-to-white contrast is (a) 50 Å and (b) 500 Å.

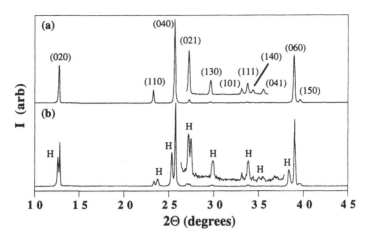

Figure 4. (a) Powder XRD pattern of pure MoO_3. (b) Powder XRD pattern of pulverized surface layers cleaved from MoO_3 single crystals reacted for 10 min. at 400 °C in $MeOH/N_2$. Peaks associated with the <203> precipitates are labeled with an H.

(Cmcm) with refined lattice parameters (**a** = 3.8830 (9), **b** = 14.0538 (25), **c** = 3.7282 (16)) identical to those of the hydrogen bronze phase, H_xMoO_3, where $0.23 \le x \le 0.4$ [7]. Every peak in the XRD pattern can be indexed to either MoO_3 or H_xMoO_3. The XRD peak positions from surface layers and powders reacted between 300 °C and 350 °C are identical. Furthermore, patterns from layers with the long [001] features (as in Fig. 3b) indicate that these features are the same H_xMoO_3 phase. These results are consistent with earlier XRD studies which showed that H_xMoO_3 forms during reactions between MoO_3 powder and $MeOH/N_2$ mixtures at 200 °C [8,9]. A more complete description of H_xMoO_3 formation in MoO_3 is found in Ref. [10].

When the (010) surface is reduced in $10\%H_2/N_2$ at 400°C, it is modified in two ways. First, crystallographic shear (CS) planes nucleate along the [001] axis of the crystal within 1 min of reduction [11-14]. In AFM images, the CS planes appear as straight, 1.5 Å surface steps [5] that form to accommodate oxygen deficiency brought on by the reducing treatment. The second modification is the nucleation of voids on the (010) surface planes (see Fig. 5). Voids nucleate within about 5 min and are initially small, with diameters less than 500 Å and depths on the order of 50 Å. The voids are bounded by step loops that must have Mo atoms in reduced coordination. With increasing reaction times, the voids grow in depth and breadth and their edges show a preferred orientation, nearly parallel to <101>. It has been demonstrated that water vapor, present as an impurity or a reaction product, catalyzes void formation and growth [5]. Structural modifications were not detected with AFM on surfaces reduced below 350 °C.

When MoO_3 samples are reduced in $10\%H_2/N_2$ at 400 °C before being reacted in $MeOH/N_2$ at 330 °C, the (010) surface is modified in much the same way as the freshly cleaved surface; acicular precipitates of H_xMoO_3 nucleate along <203>. However, the precipitation reaction is favored at the voids. After 2 min at 330 °C in MeOH, H_xMoO_3 precipitates are typically found in the vicinity of voids (see Fig. 6a). In the image, an H_xMoO_3 precipitate can clearly be seen emerging from the void labeled with a "V". Additional defects with a distinct [001] character have also formed near the voids. These features appear to be similar to defects observed at higher temperatures (400 °C) during reactions with MeOH [10] and they are distinguished from the surface/CS plane intersections because the height over the defects is typically much greater (up to 20 Å) than those observed over the CS planes. In most instances, these features coincide with a CS plane, but they are only as long as the width (along [001]) of the void to which they are adjacent. If one follows a CS plane along [001], the height over the CS plane is only 1.5 Å, but adjacent to voids, the height may be increased to more than 10 Å. This swelling is not seen on

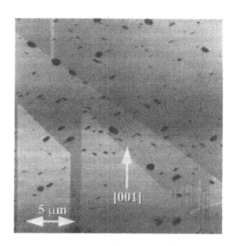

Figure 5. AFM topograph of an MoO$_3$(010) surface which was reduced in 10%H$_2$/N$_2$ for 8 min at 400°C. The black-to-white contrast in this image is 50 Å.

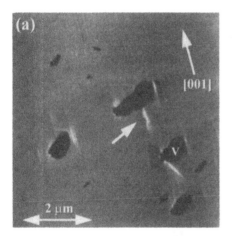

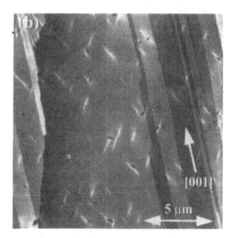

Figure 6. AFM topographs of MoO$_3$(010) surfaces reduced in 10%H$_2$/N$_2$ at 400°C for (a) 8 and (b) 4 min and then reacted at 330°C in N$_2$ saturated with MeOH (0°C) for (a) 2 and (b) 15 min. The black to white contrast in (a) is 20 Å and in (b) it is 40 Å.

surfaces which have only been reduced in 10%H$_2$/N$_2$. Finally, note that the feature indicated by the arrow in Fig. 6a shows both [001] and <203> character.

The preferential nucleation of H$_x$MoO$_3$ in the vicinity of the voids is more pronounced with longer reaction times. The H$_x$MoO$_3$ precipitates in Fig. 6b are clustered around the voids introduced during reduction. In most cases, the largest precipitates intersect a void and the void-free areas are often without precipitates. In other cases, the precipitates originate at one of the swollen [001] defects adjacent to the void. Finally, we note that similar behavior is observed when the MeOH reaction is carried out within the temperature range between 300 and 400 °C.

DISCUSSION AND CONCLUDING REMARKS

When a MeOH molecule is oxidized by MoO_3, two H are abstracted from the alcohol. According to the accepted model, the first is removed during the initial dissociative chemisorption of MeOH and the second is liberated when the surface methoxy decomposes to formaldehyde [15]. While it is usually assumed that the H react with lattice O to form H_2O, under the experimental conditions described here (in the absence of O_2), some fraction of the liberated H intercalates into the MoO_3 structure. Machiels and Sleight [16] observed that in the absence of O_2, the products formed by the reaction of MeOH and MoO_3 included dimethyl ether, hydrogen, and methane, but not formaldehyde. Therefore, methanol chemisorption still occurs in the absence of O_2, but the methoxy species are not further dehydrogenated. This suggests that the intercalating H are produced during the first step of the reaction [10].

The results presented here demonstrate that during methanol oxidation, H_xMoO_3 precipitates form preferentially at the step loops which surround the surface voids. While similar step sites are inevitably found on freshly cleaved surfaces, they have no apparent affect on the intercalation process. This is probably due to the fact that the steps on cleavage surfaces typically have single bilayer heights (7 Å) while the step loops found on pre-reduced surfaces are many times this height. There are two possible explanations for preferential formation of the H_xMoO_3 precipitates in the vicinity of the voids introduced during reduction. The first is that the voids simply act as preferential nucleation sites for H_xMoO_3. In this scenario, H enters the crystal at as yet unidentified surface sites, diffuses to the voids and, finally, precipitates heterogeneously as H_xMoO_3. The second possibility is that H actually intercalates at the void after being removed from the MeOH molecule. The voids not only provide easy steric access to the van der Waals gaps within which the H resides in the H_xMoO_3 structure, but they are also lined with Mo atoms in reduced states of coordination. Such sites have previously been linked to higher activity for the dissociative chemisorption of methanol [2]. Assuming this to be true, then we conclude that MeOH dissociatively chemisorbs at sites on the edges of the voids and as the concentration of dissolved H in the vicinity of the void increases, the H_xMoO_3 precipitate forms.

ACKNOWLEDGMENT

This work was supported by NSF YIA Grant No. DMR-9458005.

REFERENCES

[1]   J.M. Tatibouët and J.E. Germain, J. Catal., **72**, 275 (1981).
[2]   U. Chowdhry, A. Ferretti, L.E. Firment, C.J. Machiels, F. Ohuchi, A.W. Sleight and R.H. Staley, Appl. Surf. Sci., **19**, 360 (1984).
[3]   J.C. Volta and J.M. Tatibouët, J. Catal, **93**, 467 (1985).
[4]   K. Bruckman, R. Grabowski, J. Haber, A. Mazurkiewicz, J. Sloczynski, and T. Wiltowski, J. Catal., **104**, 71 (1987).
[5]   R.L. Smith and G.S. Rohrer, J. Catal., **163**, 12 (1996).
[6]   R.L. Smith and G.S. Rohrer, J. Solid State Chem., **124**, 104 (1996).
[7]   J.J. Birtill and P.G. Dickens, Mat. Res. Bul., **13**, 311 (1978).
[8]   P. Vergnon and J.M. Tatibouët, Bull. Soc. Chim. Fr., **11-12**, 455 (1980).
[9]   J. Guidot and J.E. Germain, React. Kinet. Catal. Lett., **15**, 389 (1980).
[10]   R.L. Smith and G.S. Rohrer, J. Catal., in press.
[11]   L.A. Bursill, Proc. Roy. Soc., **A311**, 267 (1969).
[12]   L.A. Bursill, W.C.T. Dowell, P. Goodman, and N. Tate, Acta Cryst., **A34**, 296 (1974).
[13]   P.L. Gai, Phil. Mag. A, **43**, 841 (1981).
[14]   P.L. Gai-Boyes, Catal. Rev.-Sci. Eng., **34**, 1 (1992).
[15]   W.E. Farneth, F. Ohuchi, R.H. Staley, U. Chowdhry, and A.W. Sleight, J. Phys. Chem., **89**, 2493 (1985).
[16]   C.J. Machiels and A.W. Sleight in <u>Proceedings of the 4th International Conference on the Chemistry and Uses of Molybdenum</u> (H.F. Barry and P.C.H. Mitchell, Eds.), p. 411. Climax Molybdenum Co., Ann Arbor, MI, 1982.

Catalytic Decomposition of Perfluorocompounds

S. KANNO*, S. IKEDA*, H. YAMASHITA*, S. AZUHATA*, K. IRIE** and S.TAMATA**
*Hitachi Research Laboratory, Hitachi, Ltd.,7-1-1 Omika-cho, Hitachi-shi Ibaraki-ken
319-12, Japan
**Hitachi works, Hitachi, Ltd., 3-1-1 Saiwai-cho, Hitachi-shi Ibaraki-ken 317, Japan

ABSTRACT

It is becoming increasingly important to decompose PFCs (Perfluorocompounds), which are powerful greenhouse gases. The process of catalytic decomposition is expected to be effective in operating at lower temperatures if catalysts of high activity and durability are developed. The decomposition activities of PFC with H_2O was investigated using several catalysts. It was found that PFC decomposition activity was related to the reactivity of fluorine and the formation of mixed oxides. Using the catalyst we have developed, CF_4 was decomposed with conversion above 99.9% over 1988h. Furthermore, another PFC was decomposed over the catalyst. The order of the conversion was CHF_3 > CF_4 > C_2F_6. The catalytic decomposition system can use the treatment of PFCs in dry etch process waste gas.

INTRODUCTION

PFCs (Perfluorocompounds), which are used in dry etch processes and CVD chamber cleaning operations in the semiconductor manufacturing industry, have the following properties : (1) They are very stable and have long atmospheric lifetimes; and (2) They strongly absorb infrared rays (The absorption coefficient of CF_4 is 6,500 times as potent as that of CO_2.). For these reasons, PFC gas causes global warming [1]. The amount of PFC consumption has increased year by year. The reduction of PFC emissions to the environment is an important subject. The U.S. Environmental Protection Agency (EPA) and the U.S. Semiconductor Industry Association (SIA) are investigating inhibition methods of reducing PFC emission [2]. Semiconductor industries are trying to reduce their emission using several methods : process optimization, alternative chemicals, recovery and recycle, and decomposition. Decomposition methods such as combustion, chemical adsorption and so on are currently being studied. The process of catalytic decomposition is expected to be effective in operating at lower temperatures and in small spaces if catalysts of high activity and durability are developed. In this paper, the PFC decomposition activity and durability of several catalysts was investigated.

EXPERIMENTAL

Catalyst preparation

Catalysts were prepared from commercial metal oxides having a diameter of 2-4 mm. These particles were pulverized and sieved to a range of 0.5-1 mm. Then, the PFC decomposition activity was investigated. Binary oxide catalysts were prepared from the powder of raw materials and several metal nitrates or metal precursors. The slurries were dried and calcined in air. The calcined catalysts were pulverized and sieved to a range of 0.5-1 mm and the activity was investigated.

Measurement of activity

The experimental apparatus used to measure the PFC decomposition activity of catalysts is shown in Fig.1. PFC decomposition activities were measured by a conventional fixed bed flow reactor. A gaseous mixture of PFC (C_2F_6, CF_4, CHF_3), H_2O and Air(bal.) was fed with a space velocity (SV) of 1,000-3,000h^{-1}. The PFC concentration changed within a range of 0.5-3 vol%. The volume of H_2O (water vapor) was 50 times as large as that of PFC.

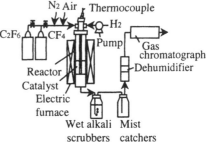

Fig.1 Experimental apparatus to measure the PFC decomposition activities of catalysts.

The concentration of PFC was measured with a gas chromatograph using either FID or TCD detectors and conversion rates were calculated from the concentration differences.

Catalyst analyses

The X-ray diffraction patterns were measured with a Rigaku Denki RU-200 X-ray analyzer.
Adsorption isotherms of nitrogen on catalysts were measured by a volumetric method using a Shimazu Accusorb-2100. The specific surface area was calculated using the B.E.T. equation from the nitrogen adsorption isotherms.

RESULTS

PFC decomposition reaction

The change in the free energy (ΔG) of representative CF_4 decomposition reactions was calculated from thermodynamic data at 1000K [3]. The thermal decomposition (Eq.(1)) and the combustion (Eq.(2)) are impractical because of the unfavorable thermodynamic equilibrium. On the other hand, it is thermodynamically possible to decompose CF_4 in the presence of such compounds as H_2O, H_2, CH_4 and O_2, because the ΔG of schemes (3), (4) and (5) show negative values. Though the methane additive reaction mostly proceeds in these reactions, the hydrolysis reaction is the most practical from the point of safety and handling. We therefore selected the hydrolysis reaction as the method to follow.

$$CF_4 = C+2F_2 \qquad\qquad \Delta G=781 \quad (kJ/mol) \qquad (1)$$
$$CF_4+O_2 = CO_2+2F_2 \qquad \Delta G=385 \quad (kJ/mol) \qquad (2)$$
$$CF_4+2H_2O = CO_2+4HF \qquad \Delta G=-343 \quad (kJ/mol) \qquad (3)$$
$$CF_4+4H_2 = CH_4+4HF \qquad \Delta G=-313 \quad (kJ/mol) \qquad (4)$$
$$CF_4+CH_4+2O_2 = 2CO_2+4HF \qquad \Delta G=-1144 \,(kJ/mol) \qquad (5)$$

PFC decomposition activity

At first, we attempted to decompose C_2F_6 over a TiO_2-based catalyst, which is known to promote the decomposition of CFCs (Chlorofluorocarbons)[4]. Though PFC does not contain any chlorine, it has the same type of halogenated carbons as CFCs. The conversion of C_2F_6 is shown in Fig.2.

Though CFC12 was easily decomposed, C_2F_6 conversion was only about 10% at 673K-1173K. For this reason, it was considered as follows.

The properties of fluorocompounds such as C_2F_6 depend on the number of fluorine atoms contained in the molecule. The van der Waals radius of a fluorine atom is 1.47nm (This value is 1.2 times as big as that of a hydrogen atom (1.20nm)). Therefore steric hindrance of adjacent functional groups in the molecule does not occur and PFCs are chemically stable. Furthermore, the C-F binding energy is bigger than other halogen bonds (ex. C-Cl bond). The binding energy increases as the number of fluorine atoms increases [5]. Because of this property, PFCs have a large intramolecular force; i.e. PFCs have a small intermolecular force and small adsorption ability. The surface of a CFC decomposition catalyst has poor reactivity towards fluorine and the catalyst tends to maintain the decomposition activity. We considered that a component which has high adsorption capacity for fluorine atoms was required in the PFC decomposition catalyst. The relationship between the fluorination enthalpies of several metals at 1100K and C_2F_6 decomposition over those metal-oxide catalysts at 1073K is shown in Fig.3. The value of ΔH was recalculated to the value of the M-F single bond. Though a relationship was not found, the order of the activity is as follows : $Al_2O_3 > TiO_2 > ZrO_2 > SiO_2$. It was assumed that Al_2O_3 formed a more stable M-F bond than the other catalysts.

In order to increase the decomposition activity, another metal oxide was added to the Al_2O_3. The conversion of C_2F_6 over several binary catalysts was found to exhibit a significant increase. XRD analyses of the catalysts showed that high decomposition activity was obtained when mixed oxides of Al_2O_3 were formed. The activities of the catalysts which did not form a mixed oxides were low. From these results, it is considered that the formation of a mixed oxides of Al_2O_3 is effective in increasing activity. The change in the free energy (ΔG) of the mixed oxides formation reaction was calculated. The relationship between the change in free energy ΔG at 1000K and the conversion of C_2F_6 at 973K is shown in Fig.4. A catalyst, which has a more negative value of ΔG, showed higher C_2F_6 conversion. These results showed the effect of mixed oxides.

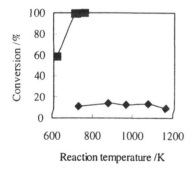

Fig.2 PFC conversion over a TiO_2-based catalyst.
Decomposition conditions :
CFC12 or C_2F_6=3%, H_2O=15%, Air=balance, SV=3,000h^{-1}.
■; the decomposition of CFC12
◆;the decomposition of C_2F_6

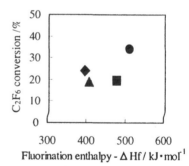

Fig.3 Relationship between fluorination enthalpy and C_2F_6 conversion.
Decomposition conditions :
C_2F_6=0.5%, H_2O=25%, Air=balance, SV=3,000h^{-1}, Cat.temp.=1073K.
● ; Al_2O_3, ■ ; ZrO_2, ▲ ; SiO_2, ■ ; TiO_2

Durability of catalysts

It is important in practical use that the catalyst have long lifetime. The durability test of CF₄ was investigated over P-100 catalyst. The CF₄ conversion at 2000h is shown in Fig.5. Initial conversion was 99.9% and this conversion level above 99.9% remained during 1988h and showed satisfactory durability. The specific surface area was decreased from $141m^2/g$ to $41m^2/g$(after 1988h). From these results, it is clear that the change in the specific surface area did not influence CF₄ conversion. It is necessary to clarify the effects of changes in catalyst properties on PFC decomposition.

Decomposition of PFCs

CF₄, C₂F₆ and CHF₃ have different numbers of fluorine atoms. To confirm the relationship between PFC decomposition activity and the number of fluorine atoms, the decomposition of several PFCs was carried out over P-100. The temperature dependence of PFC conversion is shown in Fig.6. The order of reactivity for the decomposition was as follows : CHF₃ > CF₄ > C₂F₆. The decomposition of PFCs became more difficult as the number of fluorine atoms in the PFC molecule increased. The C-F binding energy in halomethanes (CHF₃, CF₄) is bigger as the number of fluorine atoms is increased[5]. Though the C-F binding energy in C₂F₆ is smaller than that in halomethanes[5], C₂F₆ has more C-F bonds than CF₄. Therefore, we consider that PFC decomposition activity is related to the number of C-F bonds.

Decomposition apparatus

We produced a prototype apparatus of "Catalytic PFC decomposition apparatus (C-600P)". The PFC decomposition system is shown in Fig.7. This apparatus can treat 60L of waste gas, which includes 0.5vol% PFC, per minute. The size was 940mmL×500mmW×1950mmH. By adopting

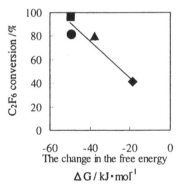

Fig.4 Relationship between the change in the free energy of the mixed oxides formation reaction of Al₂O₃ and C₂F₆ conversion. Decomposition conditions: C₂F₆=0.5%, H₂O=25%, Air=balance, SV=2,000h⁻¹, Cat.temp.=973K.

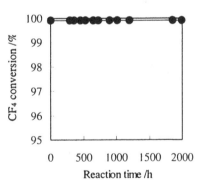

Fig.5 Durability of P-100 catalyst in the decomposition of CF₄.
Decomposition conditions: CF₄=0.5%, H₂O=25%, Air=balance, SV=1,000h⁻¹, Cat.temp.=973K.
● ; P-100

the catalyst method, the system is simple and compact. The weight is about 300kg. Waste gas which includes PFC and H_2O is heated by a heater and an electric furnace. The gas is fed to a catalyst and PFC reacts with H_2O. The decomposition gas is quenched immediately by use of water in a cooling room and a scrubber. HF is removed from the gas. This apparatus has many advantages. There is less exhaust gas and a reduction in energy is possible because the reaction temperature is low. The apparatus general view is shown in Fig.8. We have a plan to demonstrate the ability of this apparatus at a semiconductor factory.

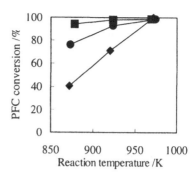

Fig. 6 The conversion of several PFCs over P-100 catalyst.
Decomposition conditions :
PFC=0.5%, H_2O=25%, Air=balance,
SV=1,000h^{-1}.
■;CHF₃, ●;CF₄, ◆;C₂F₆,

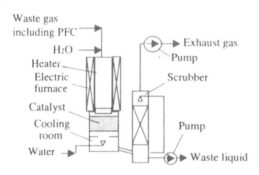

Fig.7 PFC decomposition system.

Fig.8 The apparatus general view.

CONCLUSION

For the purpose of developing a catalytic PFC decomposition system, we examined the PFC decomposition of various catalysts and reached the following conclusions.
1. PFCs (CF₄, C₂F₆ and CHF₃) are decomposed over catalysts in the presence of water vapor.
2. Reactivity with fluorine and the formation of mixed oxides influenced PFC decomposition activity.
3. Using a catalyst which we have developed, CF₄ was decomposed with conversion above 99.9% at 973K over 1988h.

REFERENCES

1. Elizabeth A. Dutrow in PFC Technical Update, edited by SEMI (SEMICON WEST 96, SEMI Technical Publication, 1996) p.1.
2. Walter F.Worth in 4th International Environment, Safety & Health Conference of The Semiconductor Industry, 1997
3. M.W.Chase, Jr., C.A. Davles, J.R.Downey. Jr., D.J.Frurip, R.A.McDonald, and A.N.Syverud., JANAF Thermochemical Tables 3rd ed., (The national Bureau of Standards Publishers, Gaitherburg, 1985)
4. S.Kanno, T.Arato, A.Kato, H.Yamashita, S.Azuhata and S.Tamata, Nippon Kagaku kaishi **2** 129(1996).
5. G.Glocker, J.Phys.Chem. **63**, 828 (1959).

REDOX BEHAVIOR BELOW 1000K OF Pt-IMPREGNATED CeO_2-ZrO_2 SOLID SOLUTIONS: AN IN-SITU NEUTRON DIFFRACTION STUDY

C.-K. Loong[1], S. M. Short[1], M. Ozawa[2], and S. Suzuki[2]

[1]Argonne National Laboratory, Argonne, IL 60439, U. S. A.
[2]Nagoya Institute of Technology, Tajimi, Gifu, 507, Japan.

ABSTRACT

The $Ce^{3+} \leftrightarrow Ce^{4+}$ redox process in automotive three-way catalysts such as Ce-ZrO_2/Pt provides an essential mechanism to oxygen storage/release under dynamic air-to-fuel ratio cycling. Such a function requires a metal-support interaction which is not completely understood. We have carried out an *in-situ* neutron powder diffraction study to monitor the crystal structures (a mixture of a major tetragonal and a minor monoclinic phase) of 10mol% Ce-doped ZrO_2 with and without Pt (1wt%) impregnation under oxidizing and reducing conditions over the temperature range of 25°-700° C. The samples were heated first in flowing 2%O_2/Ar from room temperature to 400° C and then in 1%CO/Ar to about 700° C. A discontinued increase of the tetragonal unit-cell volume, a decrease of tetragonality (c/a), and a change of color from light yellow to gray when changing from oxidizing to reducing atmosphere were observed only in the sample containing Pt. This result supports the model which assumes the formation of oxygen vacancies initially near the Pt atoms. As more Ce ions are reduced from 4+ to 3+ oxidation states at high temperatures, oxygen vacancies migrate to the bulk of the oxide particles.

INTRODUCTION

The present study is motivated by the need of better support materials for noble metals in the three-way catalytic converters for automotive exhaust gas control, but it is also relevant to a wide variety of applications such as oxygen sensor technology and petroleum reforming. The catalytic systems of interest are noble metals (e. g., Pt) dispersed on cerium doped zirconia fine powders. The exact role of ceria in the promotion of simultaneous NO reduction and CO oxidation under dynamic air-to-fuel cycling of the engine is still not understood. The microstructure of the CeO_2-ZrO_2 solid solutions, and the formation of oxygen vacancies in the crystalline lattice in conjunction with the redox process of the Ce ions are thought to be the important factors.[1] Recently we have carried out a series of studies of high-surface-area rare-earth modified zirconia powders using the method of neutron scattering. The microstructure in terms of primary-particle size, surface area, porosity and fractal aggregates of the oxide supports have been characterized by small-angle scattering.[2] The global crystal phases and the nature of short-range oxygen defects in $Ln_{0.1}Zr_{0.9}O_{2-x}$ (Ln = La, Ce and Nd) were investigated by neutron powder diffraction.[3] In the case of La(III)- and Nd(III)-doped ZrO_2, a real-space correlation function, obtained from a Fourier transform of the filtered residual diffuse scattering, showed evidence of static, oxygen vacancy-induced atomic displacements along the pseudocubic <111> and other directions. No such defects were found in Ce-ZrO_2 because Ce has the same oxidation state (4+) as Zr in the solid solution.

It is well know that high-surface-area ceria may undergo reduction/oxidation cycles according to the reaction $2CeO_2 \leftrightarrow Ce_2O_3 + 1/2O_2$ whereby the Ce valence oscillates between the 4+ and 3+ states. Temperature programmed reduction experiments on pure ceria in hydrogen revealed two reduction peaks at about 500° and 800° C corresponding to reduction of Ce ions on the surface and in the bulk, respectively.[1, 4] Magnetic susceptibility measurements of heat-treated ceria under CO showed a similar result.[5] It is also found that adding alkaline earth and transition-metal oxides to CeO_2 enhances the thermal stability of ceria against sintering at high temperatures. Dispersion of platinum group metals on ceria-containing oxide powders dramatically increases the oxygen storage/release capability of ceria in the simultaneous conversion of CO and NOx to CO_2 (oxidation) and N_2 (reduction). The interaction of platinum with ceria may be rationalized by a model which involves associative adsorption of CO on Pt/CeO_2 followed by a reaction to produce CO_2 and an oxygen vacancy.[1] During this process oxygen is released and Ce 4+ → 3+ transition occurs. Since the difference in the ionic radius of Ce^{4+} (0.80 Å) and Ce^{3+} (1.01 Å) is substantial, a change of the lattice cell volume of the oxide is expected if the transition involves Ce ions in the bulk. In this paper we present the results of an in-situ neutron diffraction study of the redox behavior of a 10 mol% CeO_2-ZrO_2 solid solution with and without Pt over the temperature range of 25°-700° C under an oxidizing or reducing atmosphere.

EXPERIMENTAL DETAILS

The $Ce_{0.1}Zr_{0.9}O_2$ powders, with and without impregnated Pt (1 wt%), were prepared by a coprecipitation method as described elsewhere previously.[6] The fresh powders that were subjected to heating in air for 3-5h showed a BET surface area of ~26 m^2/g and an average particle size of ~21 nm.[2]

The neutron diffraction experiments were carried out using the Special Environment Powder Diffractometer at the Intense Pulsed Neutron Source of Argonne National Laboratory. A powder was compacted to form pellets (cross-sectional diameter of 11 mm) which were stacked to form a column of about 50 mm tall inside the sample tube of a furnace. A flowing-gas (~100-150 ml/min) sample environment was maintained throughout the experiments at all temperatures. The data were collected by detectors situated at mean scattering angles of ±90° for which a resolution of $\Delta d/d = 0.54\%$ can be achieved (d is the atomic plane spacing). By virtue of the highly collimated neutron entrance and exit beams in conjunction with a 90°-scattering geometry, the detectors admit solely Bragg-scattering intensity from the sample. The sample temperature, monitored by thermocouples above and below the sample, were controlled at a selected temperature to within 15°C in all runs. The diffraction data were collected every hour in the following manner for both samples: heating from 25° to 400° C in flowing $2\%O_2/Ar$ gas followed by continued heating from 400° to 700°C in flowing $1\%CO/Ar$ gas. The temperature was stepped at an interview of 50° C and a 20-min wait prior to the data collection was allowed for thermal equilibration at each temperature. At 400°, 500°, 600° and 700° C multiple 1h-data sets were collected. The data were analyzed by the Rietveld refinement technique using the Generalized Structural Analysis System (GSAS) computer code.[7]

RESULTS AND DISCUSSION

Pure ZrO_2 has three polymorphs: a cubic fluorite structure (space group Fm3m) above 2640 K, a tetragonal structure (P4$_2$/nmc) between 1400 and 2640 K, and a monoclinic structure (P2$_1$/c) below 1400 K. The tetragonal and cubic phases over a wide range of temperatures can be stabilized by a small amount of rare-earth solutes. The exact crystal phases and microstructure of rare-earth modified zirconia powders, however, may vary depending on the rare-earth dopant and the processing route.[8] The present samples, regardless of the presence of Pt metal, compose of a major tetragonal phase (74±2 mol%) and a minor monoclinic phase (26±4 mol%). The relative phase fraction did not change with respect to temperature. Fig. 1 shows a diffraction pattern for Ce-ZrO$_2$/Pt at 400° C in a flowing CO/Ar environment. The general features in the intensity profile are typical to all runs. In particular, the broad peaks and non-Gaussian line shapes indicate the small crystalline grain size and the presence of microstructure. A detailed analysis of the diffraction data of Ce-, La- and Nd-doped zirconia powders has been given in another publication.[3] Contributions to the intensity profiles from the dispersed Pt particles in the Ce-ZrO$_2$/Pt sample or oxygen-vacancy defects in the reduced samples are too small to be analyzed quantitatively by the Rietveld refinements. All the refinements were made using a nominal sample stoichiometry of $Ce_{0.1}Zr_{0.9}O_2$.

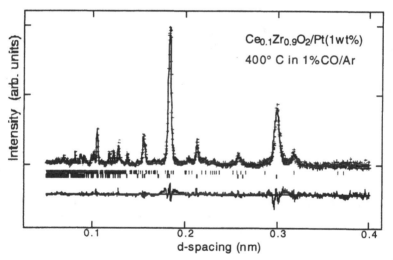

Figure 1. Rietveld profile fit in the 0.04-0.4 nm region of d-spacing for the $Ce_{0.1}Zr_{0.9}O_2$ powder impregnated with 1 wt% of Pt metal at 400° C under a flowing 1%Co/Ar gas environment. The symbols are the observed, background subtracted intensities. The solid line represents the calculated crystalline intensities. The tick marks indicate the positions of the Bragg reflections of the monoclinic (top row) and tetragonal (bottow row) phases. The difference between the observed and calculated intensities is shown at the bottom of the figure.

Fig. 2 shows the lattice parameters of the tetragonal phase of Ce-ZrO$_2$ and Ce-ZrO$_2$/Pt as a function of temperature and sample environment. In the case of Ce-ZrO$_2$ (Fig. 2a) both a and c increase linearly with increasing temperature regardless of the oxidizing atmosphere below 400° C or the reducing atmosphere above 400° C. The small differences in the lattice parameters at 400° C arose from systematic errors in the experiment because the run with O$_2$/Ar was conducted at a different time from that with CO/Ar. In the case of Ce-ZrO$_2$/Pt the two runs were conducted in sequence without removing the sample. After the completion of heating the Ce-ZrO$_2$/Pt sample up to 400° C under O$_2$/Ar, the sample environment was evacuated and then CO/Ar gas was fed into the furnace tube. Fig. 2b clearly shows an increase of the basal plane lattice parameter a at 400° C after CO/Ar gas was introduced into the system. The lattice parameter c, on the other hand, show little or no increase over the 6h period at 400° C under flowing CO/Ar. An examination of the tetragonality (c/a) of the lattice showed that under oxidizing conditions c/a increased (from 1.018 to 1.019) with temperature from 25° to 400° C for both samples. Under reducing conditions c/a decreased with increasing temperature for both samples, but the decrease was larger for the Ce-ZrO$_2$/Pt sample.

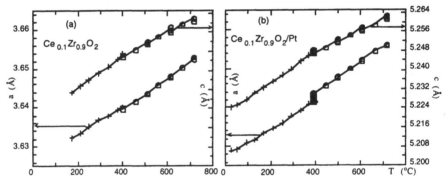

Figure 2. The lattice parameters of tetragonal Ce$_{0.1}$Zr$_{0.9}$O$_2$ (a) and Ce$_{0.1}$Zr$_{0.9}$O$_2$/Pt (b) versus temperature. The samples were first heated in 2%O$_2$/Ar (+) and then switched to 1%CO/Ar and continued heating to 700° C (o). Each data point represents 1h data collection.

The unit-cell volumes of the tetragonal and monoclinic phases for Ce-ZrO$_2$ and Ce-ZrO$_2$/Pt are shown in Fig. 3. There is an volume increase of ~0.25% only in the Ce-ZrO$_2$/Pt sample under reducing condition. The effect of the redox of Ce on the monoclinic phase is less clear, perhaps due to the relatively large uncertainties in the refinement of this minor phase. No obvious increase of the lattice volume attributable to the reducing environment is evident from the data. From the Rietveld refinements the volume of the monoclinic phase for the Ce-ZrO$_2$/Pt sample is slightly larger (~0.15%) than that for the Ce-ZrO$_2$ sample throughout the whole temperature range. The origin of this difference is not yet understood. More measurements of Ce-ZrO$_2$ solid solutions over a wide range of compositions have been planned for the future.

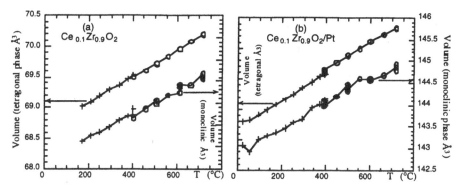

Figure 3. The unit-cell volumes of the tetragonal and monoclinic phases of $Ce_{0.1}Zr_{0.9}O_2$ (a) and $Ce_{0.1}Zr_{0.9}O_2$/Pt (b) versus temperature. The samples were first heated in $2\%O_2$/Ar (+) and then switched to 1%CO/Ar and continued heating to 700° C (o). Each data point represents 1h data collection.

CONCLUSION

The lattice parameters of the tetragonal and monoclinic phases of $Ce_{0.1}Zr_{0.9}O_2$ with and without Pt impregnation increase linearly with increasing temperature over the 25°-700° C range. From Fig. 1 the thermal expansion coefficients for the tetragonal phase at ~300° C are $\alpha_a = 9.5 \times 10^{-6}$ /°C and $\alpha_c = 13.1 \times 10^{-6}$ /°C. These values are comparable with those reported for Sc, In or Yb stabilized zirconia.[9] Switching from an oxidizing atmosphere for 25°-400°C to a reducing one for 400°-700° C in heating the Ce-ZrO₂ sample does not induce an anomaly in the thermal expansion. Since the observed d-spacing reflects the long-range order structure in the crystalline grains, the lack of anomaly implies that the reduction of Ce ions, if any, occurs mainly on the surfaces the oxide particles. If Ce-ZrO₂ is impregnated with Pt metal, on the other hand, an increase of the axial lattice parameter (c) of the tetragonal phase when switching from oxidizing to reducing conditions was observed. Consequently, the tetragonality decreases. This suggests a gradual (over several hours) conversion of the 4+ to 3+ state of the Ce ions. This process probably occurs initially in the interfacial region between the oxide and Pt particles and then migrates to the bulk of the oxide lattice. This phenomenon can also be seen visually from the change of color (from light yellow before to gray after reduction) only for the Pt-impregnated sample. The original color can be recovered by heating the gray sample under an O_2-rich atmosphere. The origin for the color change in zirconia resulted from heat treatments was a subject of debate previously.[10-12] In the present case, it is undoubtedly caused by the redox process of Ce ions.

ACKNOWLEDGMENTS

Work performed at Argonne National Laboratory is supported by the U. S. DOE-BES under Contract No. W-31-109-ENG-38.

REFERENCES

1. B. Harrison, A. F. Diwell, and C. Hallett, Platinum Metals Rev. **32**, 73 (1988).
2. C.-K. Loong, P. Thiyagarajan, J. Richardson, J. W., M. Ozawa, and S. Suzuki, J. Catal. **171**, 498 (1997).
3. C.-K. Loong, J. W. Richardson, Jr., and M. Ozawa, J. Catal. **157**, 636 (1995).
4. A. Trovarelli, Catal. Rev. - Sci. Eng. **38**, 439 (1996).
5. A. Badri, J. Lamotte, J. C. Lavalley, A. Laachir, V. Perrichom, O. Touret, G. N. Sauvion, and E. Quemere, Eur. J. Solid State Inorg. Chem. **28**, 445 (1991).
6. M. Ozawa, M. Kimura, and A.Isogai, J. Alloys Compounds **193**, 73 (1993).
7. A. C. Larson and R. B. von Dreele, (Los Alamos National Laboratory, 1985) Report LAUR 86-748.
8. M. Ozawa and M. Kimura, J. Less-Common Metals **171**, 195 (1991).
9. T.-S. Sheu, J. Am. Ceram. Soc. **76**, 1772 (1993).
10. J. Soria and J. S. Moya, J. Am. Ceram. Soc. **74**, 1747 (1991).
11. R. W. Rice, J. Am. Ceram. Soc. **74**, 1746 (1991).
12. J. S. Moya, R. Moreno, J. Requena, and J. Soria, J. Am. Ceram. Soc. **71**, C479 (1988).

NEW TYPE OF ADVANCED CATALYTIC MATERIAL BASED UPON ALUMINA EPITAXIALLY GROWN ONTO THIN ALUMINA FOIL

S.F.TIKHOV*, G.V.CHERNYKH*, V. A. SADYKOV*, A. N. SALANOV*, S. V. TSYBULYA*,G.M. ALIKINA*, V. F. LYSOV**.
* Boreskov Institute of Catalysis SB RAN, Av.Lavrentieva 5, Novosibirsk, 630090, sadykov@catalysis.nsk.su
** Institute of Applied Physics, Arbuzova St. 1/1, Novosibirsk, 630117; RUSSIA

ABSTRACT

Formation of planar layered Al_2O_3/Al composites via aluminum foil anodic-spark oxidation was studied by using SEM and XRD. A lateral growth of the primary non-porous alumina islands was shown to proceed to an extent where the entire surface was covered, and this layer is responsible for determining the high thermal stability of composite. A washcoating by an alumina suspension substantially increases pore volume and water absorption capacity. Catalysts based on the platinum group metals and transition metal oxides exhibit high activity for the reactions of CO and hydrocarbon oxidations and the selective reduction of nitrogen oxides by hydrocarbons.

INTRODUCTION

Metal-oxide composites are of great interest for catalysis due to their high mechanical strength and thermal conductivity exceeding those for traditional oxide supports. Generally, such composite supports are obtained by washcoating, plasma spraying and other methods [1,2]. However, for these systems, oxidic layer spallation is often a problem due to a difference in the thermal expansion coefficients, lattice crystallographic misfit and poor adhesion [3]. In an attempt to overcome these problems, anodic oxidation of aluminum to form the primary oxidic layer firmly attached to metal seems promising approach. This work presents the key features of the oxidic layer formation via aluminum foil anodic-spark oxidation along with the catalytic properties of selected metal and metal oxide particles catalysts on these supports.

EXPERIMENTAL

Aluminum foil with a thickness of 0.3 mm was anodic-spark oxidized as in [4] at a constant current density with a subsequent air calcination at 600-900 °C. An amount of the primary oxide thus formed was estimated gravimetrically. A second preparation stage includes washcoating of the secondary alumina layer from suspensions based upon the hydrated alumina. In order to increase the thermal stability of the oxidic layer, rare-earth oxides were introduced by impregnation from nitrates solutions. Active components containing transition metal oxides and platinum group metals were supported either via impregnation or by washcoating from suspensions of catalysts with hydrated alumina. The honeycomb structure was shaped by stacking flat and corrugated sheets of foil and winding them into the Arkhimed spiral. In this way, triangular channels ~ 3x3x3 mm were formed. The catalysts have an apparent density of ~ 1g/ml. Catalytic activity for propane oxidation was determined in an integral flow reactor at 500 °C, GHSV = 50,000/h and the initial reaction mixture composition (vol.%): 0.13 C_3H_6 + 1.0 O_2 + 2.0 H_2O + He. The catalytic activity in the reaction of butane (steady state concentration of Bu in air 0.2 vol.%) and CO oxidation (initial concentration - 1.0 vol.% of CO in air) was determined in a batch-flow kinetic system. The process of the primary and secondary oxide formation was studied by using SEM

71

(BS-350 "Tesla") and XRD (URD-6 with monochromatic CuK_α radiation). Specific surface area was studied by adsorption of Ar.

RESULTS AND DISCUSSION

Formation of the primary oxidic layer

According to SEM examination, the surface of the initial foil was found to be rather smooth with the traces of the rolling action being clearly visible (Fig. 1). For the Al foil, only 200 and 220 X-ray peaks were observed while the 111, reflection the most strong for the case of powdered Al, was absent here. This finding suggests a strong orientation of the Al (100) and (110) type planes parallel to the foil surface.

At the very beginning of oxidation, the foil surface exhibited a complex texture: small porous particles were located between thick (up to 5 μm) and compact oxidic islands. According to X-ray diffraction data, the oxidic layer was mainly formed by a low-temperature form of γ-alumina (the typical 220 and 440 reflexes being observed). Some orientation of the alumina particles with respect to underlying textured aluminum foil could be thus inferred. As admixtures, aluminum phosphates were present as evidenced by diffraction peaks corresponding to 4.17 and 4.47 Å lattice spacings of different phosphate phases. These admixtures can be formed as a result of a capture of phosphate anions by the freshly formed anodic layer, as was observed earlier in a number of cases [6]. At a higher magnification, cylindrical pores were visible within small particles (Fig. 1 c, d). Such pores could be the current channels where microarcs were generated [4]. As the time of oxidation increased, dense oxidic layer proceeded to cover all the surface of aluminum foil. This layer was divided into segments whose boundaries were saturated by cracks and pores (Fig. 1 e, f). At the final stage of oxidation when electrolysis voltage approached a constant value (cf. inset in the Table 1), cracks seemed to be mainly healed (Fig.1 g, h). During the course of oxidation, the thickness of the compact oxidic particles tended to increase: from 4-5 to 8-10.5 μm. This behavior agrees well with the model of the lateral growth of oxidic layer nucleated as islands at some preferable sites. When the foil surface was covered completely by the oxidic layer, the rate of the anodic spark oxidation delined sharply.

In addition to cylindrical pores normal to the foil surface, SEM observation for oxidic layer cleaved by the foil bending revealed the existence of large pores along the surface intersecting normal ones (Fig. 2). These types of pores can appear due to oxide recrystallization under the effect of plasma located at the metal-oxide interface and generated by electron break-through in the course of anodic spark oxidation [4]. The absence of detectable amounts of high-temperature modifications of alumina (alpha etc) could be assigned to very high cooling rates in the highly non-equilibrium spark conditions favoring formation of metastable γ-alumina.

Table 1 lists the main features of the traditional anodic and anodic-spark oxidation of aluminum carried out in comparable conditions, along with some data on the properties of the anodic layers thus obtained. As follows from these data, for anodic-spark oxidation, the rate of the oxidic layer growth as well as the limiting thickness obtained are much higher, while pore density is much lower. As a result, despite bigger pore radii for the case of anodic-spark oxidation, the average value of the foil surface not protected by oxidic layer is lower here.

Thermal stability of composites

One of the most intriguing properties of supports and catalysts based upon the anodic-spark oxidized aluminum foil was their high thermal stability: they retain mechanical strength, integrity and catalytic activity even after heating to temperatures exceeding that of the aluminum

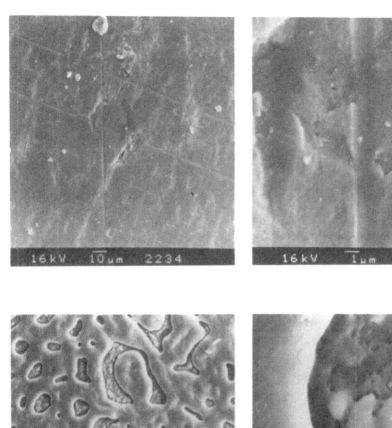

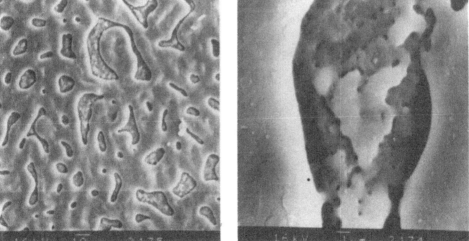

Fig 1.. SEM micrographs of aluminum foil after different time of ASO (min):.
a, b – 0; c, d – 3.

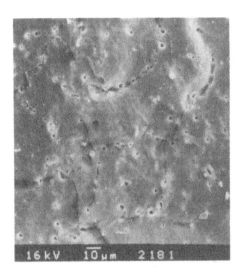 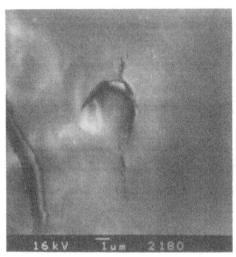

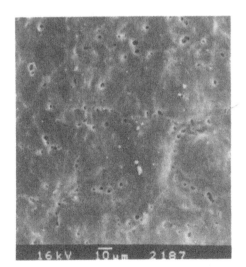 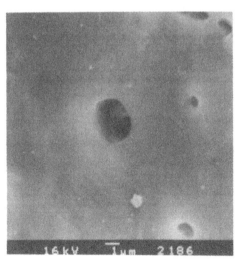

Fig 1.(Continued) SEM micrographs of aluminum foil after different time of ASO (min):.
e, f – 11; g, h – 15; b, d, f, h – higher magnification.

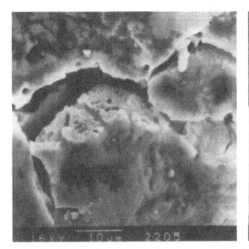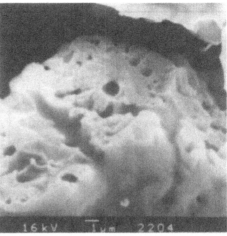

Fig.2. SEM micrographs of anodic oxidic layer after 15 min of ASO near large crack. b–higher magnification.

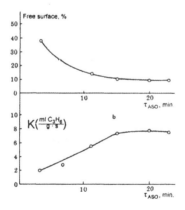

Fig.3. The aluminum foil free surface portion (a) and propane oxidation rate constant (b) vs. the time of ASO proceeding over Al_2O_3/Al -based catalysts calcined at 700 °C in air.

melting (~ 660 °C.). This indicates that molten aluminum is confined within primary oxidic layers without any leakage or dripping. After high-temperature (900 °C) calcination, the phase composition of these layers exhibited appreciabe changes: now a polycrystalline (-alumina phase predominates with only traces of the low-temperature (γ, δ, θ) phases detectable. In addition, a cubic phase with the lattice parameter a ~ 5.47 Å appears which can be assigned to aluminum phosphides [5] formed via interaction of phosphates with the molten aluminum.

Table 1. Comparison of the conventional anodic and the anodic-spark oxidation and properties of the oxide layers formed

Parameters.	Conventional anodic oxidation	Anodic-spark oxidation
Process characteristics: a) voltage vs. time	*Voltage (V) vs. Time (arb. un.) plot: 0, 100, 200, 300, 400; "Spark electrolysis", "Conventional electrolysis"*	
b) rate of foil growth (v)	$v \sim 0.1\ \mu m$	$v \sim 0.5\text{-}0.7\ \mu m$
Oxidic layer properties: a) concentration of cylindrical pores (N)	$N\ (\ 10^9$ pores/cm^2	$N\ (\ 10^5$ pores/cm^2
b) average pore radius (r)	$r \sim 0.1\text{-}0.2\ \mu m$	$r \sim 1\text{-}2\ \mu m$
c) average foil thickness (Δ)	$\Delta \sim 1\text{-}2\ \mu m$	$\Delta \sim 5\text{-}10\ \mu m$
d) surfaced covered by cylindrical pores (Q)	$Q \sim 30\text{-}40\%$	$Q \sim 5\text{-}10\%$
Reference	[6]	Our data

For the air calcined samples, stability of such a phase is rather unexpected and can be probably explained by its location within closed pores of the oxidic layer. Another specific structural feature of the composites calcined at high temperatures was a change of the aluminum grains orientation: only 220 type X-ray reflections were observed. This indicates that crystallization of the molten aluminum proceeds in a manner such orientational matching with the primary oxidic particles is maintained. SEM data revealed that all these transformations proceeded without any noticeable large-scale rearrangement of the texture and integrity of the primary oxidic layer, probably, due to a well-known topotactic character of the γ–α phase transition [7].

For the protected foil, air oxidation with a maximum 5-10% of weight gain was manifested only by formation of the volcano-shaped cones at outlets of cylindrical pores and accumulation of the dispersed oxide particles nearby. In this case, the molten aluminum and/or its vapors are transported through the pores to the surface and are subsequently oxidized by air. Quite different features were observed for a high-temperature oxidation of the non-protected foil. In this case, a thin (ca 10 Å) layer of a corundum -type oxide present at the foil surface also ensures its integrity after calcination, provided that the foil is fixed at a flat ceramic substrate, though up to 80-90% of aluminum was oxidized. SEM data have shown the oriented edge-on growth of thin corundum platelets, the most developed planes being normal to the foil surface. Such platelets were nucleated at surface steps arranged in a spiral pattern, probably decorating the outlets of screw dislocations. Here, oxidation appears to proceed via migration of Al cations from the interface along the cationic layers in corundum particles, while oxygen was incorporated at a growing edge. Hence, for unprotected foil, the oriented anisotropic growth of corundum platelets occurs without favoring formation of the compact oxidic layer at the metal-oxide interface. As a result, a pure foil is much more easily oxidized as compared with that protected by the oxidic layer grown through the anodic-spark oxidation.

Catalytic activity of composite-based systems

After deposition of the secondary alumina layer and its stabilization by rare-earth elements, the water capacity and specific surface area were increased up to 0.15- 0.20 ml/g and 5-7 m^2/g, respectively. However, despite a rather big secondary layer thickness approaching ~ 50 (m, the thermal stability of supported catalysts was found to mainly depend upon the primary oxidic layer coverage (Fig. 3).

Table 2. Shows the high activities of the developed catalysts for the butane oxidation as compared with that of traditional catalyst systems. Though these new catalysts are somewhat less thermally reaction stable, they retain a reasonably high level of activity even after calcination at temperatures exceeding the melting point of aluminum. For the CO oxidation reaction at 300 °C., Pd-based catalysts ensured conversion ranging from 99 to 97 % when the space velocity was changed from 50,000 to 300,000 h^{-1}. For Pt-Pd-Rh active component typical of car exhaust catalysts [8], CO conversion at the same temperature and very high space velocity ~ 800,000 h^{-1} reached 85%. These results are superior to those of the traditional three-way catalysts. For the reaction of NO_x selective reduction by propane, catalysts based upon copper-exchanged ZSM-5 supported on the Al_2O_2/Al composite honeycomb support using a proprietary binder [9] demonstrated NO_x conversion ~ 50% at 375 °C. and space velocity ca 50,000 h^{-1} , that is promising for practical application.

Table 2. Changes in complete butane oxidation of the catalysts after heating at 600 and 700 °C.

Catalyst (support)	Active component (wt.%)	Activity, W 10^2 (ml butane/g s) after calcination at:			
		600 °C		700 °C	
		400 °C	500 °C	400 °C	500 °C
Honeycomb (stainless steel) SHPAK-0.5	Pt (0.12)	0.3	2.1	0.2	2.0
(alumina)	Pd (0.2)	1.5	4.1	1.5	4.1
Honeycomb (Al_2O_3/Al)	Pd(0.2)	1.4	7.9	0.4	4.8
	MnO_x(2.5)	1.5	4.6	0.1	1.2

CONCLUSIONS

Catalysts based upon the anodic-spark oxidized alumina foil with a secondary washcoated oxide layer are promising for the exhaust gases clean-up provided there is no overheating. The methods for such catalyst synthesis elaborated in this work can be used to support catalytically active layers on various aluminum constructions including honeycomb supports, grids, tubes, heat exchangersetc.

REFERENCES

1. N. A. Zakarina, M. K. Yusupova in Honeycomb support and catalyst, edited by Institute of Catalysis, Novosibirsk (Proc. 2nd Workshop, 1992), p.96-100 [in Russian].
2. A. F. Feofilov, B. V. Formakovskii, A. P. Khinski, E. N. Yurchenko, Ibid, p.113-115.
3. M. Nagayama, H. Takahashi, M. Koda, J. Metal. Finish. Soc. Jap., **30**, p.438-456 (!979).
4. V. I. Chernenko, L. A. Snejko, I. I. Papanov, Preparation of washcoats by anodic-sparc electrolysis, "Khimiya", Leningrad, 1991, 150 p. [in Russian].
5. ASTM JCDS Data File 12-470.
6. V. F. Surganov, A. M. Mozalev, N. N. Mozaleva, Russian J. Appl. Chem., **70**, 267-272 (1997) [in Russian].
7. W. Krisman, P. Kurze, K.-H. Dittrich, H. G. Schneider, Crystal Res. Technol.,**19**, 973-979 (1984).
8. R.L.Keiski, M.H(rkonen, A Laht, T.Maunula, A.Savim(ki, and T.Slotte in CAPOC3 , edited by A Frennet and J.-M. Bastine, Universit(Libre de Bruxelles, Brussele (Proc.Third Intern. Cong. Catal.@ Automotive Pollution Control, 1994), v.1, p.53-63.
9. Russian Patent N2072897.

A NOVEL TIO$_2$ FILM CATALYST
— PREPARATION, PROPERTIES AND RESEARCH ON ITS PHOTOCATALYTIC OXIDIZED ACTIVITY

Y.A. CAO*, X.T. ZHANG*, L.Q. CHONG**, D.Y. WANG*, T.F. XIE*, Y. HUANG*, Z.F. CUI**, W.G. SHI*, X.J. LIU*, Z.Y. WU*, Y.B. BAI*, T.J. LI*, Y.WU***

*Department of Chemistry, Jilin University, Changchun 130023, P.R.China, yubai@mail.jlu.edu.cn
**Department of Chemistry, Changchun academy of Metallurgy and Geology, Changchun 130023, P.R.China
***Changchun Institute of Applied Chemistry, Changchun 130023, P.R.China

ABSTRACT

A new kind of TiO$_2$ film catalyst was prepared by the Plasma-enhanced chemical vapor deposition (PECVD) method. The surface photovoltaic spectroscopy (SPS) results showed that its photoresponse was extended into the visible region. Photooxidation experiments showed that this kind of TiO$_2$ film had high photocatalytic activity on degradation of phenol in aqueous solution. The influence of the thickness of TiO$_2$ film on its photocatalytic activity was also discussed.

INTRODUCTION

TiO$_2$ has been proven to be an efficient catalyst for the photodegradation of organic pollutants, owing to its high photostability and photooxidized ability[1,2,3]. However, the large band gap (3.2 eV) of TiO$_2$ also confined its ability to harness the sunlight efficiently. In recent years, many efforts have been made in order to improve its efficiency and applicability, for example, using size-quantized TiO$_2$ in order to increase the rate constant of charge transfer [4], doping TiO$_2$ in order to modify its electronic properties[5,6], and adopting composite structure to suppress the recombination of photogenerated carriers [7] etc. In the present work, we prepared a new kind of TiO$_2$ film catalyst by the well-developed PECVD method. SPS study showed that the photo response of such TiO$_2$ films was extended into the visible region, which indicated the possibility to harness visible light. Photodegradation experiments on phenol solution confirmed that high catalytic activity can be achieved. Furthermore, the effect of film thickness on the catalytic activity was also investigated.

EXPERIMENT

Preparation and characterization of TiO$_2$ thin film photocatalysts

The TiO$_2$ films were prepared on glass slide (75mm × 25mm) by the method of PECVD. In our experiment, TiCl$_4$ and O$_2$ were used as reactants, high-frequency electric field was produced by a GP300-6 high frequency plasma generator, the temperature of the glass slide was 150 ℃, and the vacuum tightness of the system was about 3×10^{-1} mmHg. By controlling the deposition time, TiO_2 films with different thickness can be prepared. Then, after calcining at 450 ℃ for half an hour, TiO_2 thin film catalysts were obtained.

The surface morphology and thickness of the resulting films were investigated by a X-650 scanning electron microscope (SEM). X-ray diffraction (XRD) patterns were recorded with a Rigaku D/max rA X-ray spectrometer. SPS spectra were measured with a laboratory-built surface photovoltaic spectrometer.

Photodegradation experiments

Photodegradation experiments were carrried out in a 70 ml pyrex glass reaction vessel with magnetic stirring. Illumination with $\lambda > 290$ nm was carried out by a 400 W high-pressure mercury lamp. The volume of phenol solution (50 ppm) was 40 ml. A glass slide coated with TiO_2 film was immersed in the middle of it. The reaction temperature was maintained at 25 ℃. During the experiment, oxygen was bubbled into the reacting mixture at a constant flux of 5 cm^2/min.

The residual concentration of phenol in solution was determined by the standard colorimetric method. For each time, 2 ml reacting solution was taken out for analysis.

RESULTS AND DISCUSSIONS

Preparation of TiO_2 thin film catalysts

The principle of PECVD is given in Ref.[8]. In this experiment, $TiCl_4$ and O_2 molecules were introduced into the reaction chamber in vapor phase, ionized by the high frequency electric field and reacted. The reaction equation thus can be expressed as follows:

$$TiCl_4 + O_2 \rightarrow TiO_2 + 2Cl_2 \tag{1}$$

In our experiment, we maintained the partition pressure of $TiCl_4$ and O_2 as 2×10^{-1} and 1×10^{-1} mmHg, respectively, only changing the reaction time, and obtained a series of TiO_2 thin film catalysts with different thicknesses. The experimental conditions and results were list in Table I. XRD patterns showed that they were anatase structure.

Table I   Experimental parameters of TiO_2 films

Sample	Depositing time (min.)	Film weight (mg)	Film thickness (μm)[a]
1[#]	10	0.20	0.14
2[#]	20	0.66	0.17
3[#]	30	0.81	0.33
4[#]	50	2.93	1.50

[a] Determined by the SEM

Fig.1 showed the SEM images of TiO$_2$ films with different deposition time, from which the formation process of the film can be concluded. In the preliminary stage of film-forming, TiO$_2$ deposited preferrably on the active sites of the glass slide and formed uniform particles. The particles grew up gradually and were connected by many little particles formed later. At that time, a loose particulate film was formed (see Fig.1a). Growing further, the larger particles in the film coagulated together gradually and a coarse but continuous film (see Fig. 1b) was formed. Keeping on reacting, the thickness of the film increased continuously, the surface of the film turned smooth gradually, and at last a dense and uniform film was formed (shown in Fig.1c). The formation process of the film indicated that the specific surface area of the film would be decreased with the film's growth, the film with different thickness may have different activity.

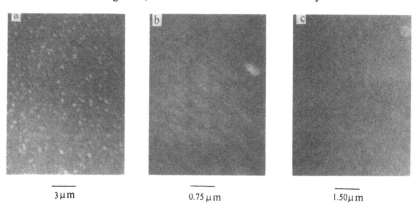

| 3µm | 0.75 µm | 1.50µm |

Fig.1 SEM images of TiO$_2$ films with different depositing time (a) 10 min. (b) 20 min. (c) 50 min.

SPS results

Fig.2a showed the SPS of TiO$_2$ particulate film. We can find that its photoresponse was broadened from 390 nm to 500 nm. When a +0.5 V electric field was applied (see Fig.2b), three distinct response bands can be distinguished which appears at 350, 390 and 430 nm respectively. According to the principle of SPS [9], the response band at 350 nm can be assigned to the transition from valence band (O$_{2p}$) to conduction band(Ti$_{3d}$), and its thresold was at 385 nm, corresponding to the band gap of 3.22eV. The response bands at 390 nm and 430 nm were contributed by the surface states, which can be assigned to the transition betw

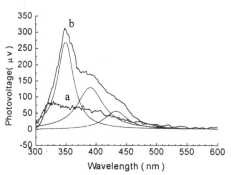

Fig.2 SPS of TiO$_2$ film prepared by the PECVD method at (a) 0.0 V(×10) and (b) +0.5 V electric field

surface state levels. As we know, photogenerated holes play a key role in the photooxidized reactions, such extended photoresponse implied a high photocatalytic activity for this kind of film catalysts.

Results of phenol photodegradation

Blank experiment showed that phenol degraded slowly under illumination without TiO$_2$ films. When a TiO$_2$ film was introduced, the apparent degradation occurred. Fig.3 showed the photodegradation results of TiO$_2$ film with different thickness. We can find that the average degradation rate changed with the thickness of film. The relation between them was shown in Fig.4 , which indicated that an optimal thickness may exist.

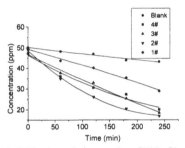

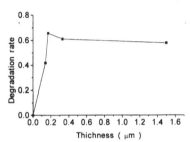

Fig.3 The degradation results of TiO$_2$ films (1$^{\#}$,2$^{\#}$,3$^{\#}$,4$^{\#}$) on phenol

Fig.4 Relation curve of catalytic activity vs film thickness

In Fig.5, a first-order plot for the kinetics was reported for the experimental data shown in Fig.3. The data could reasonably be fitted to a straight line, and thus the decomposition rate constants and t$_{1/2}$ for the photodegradation reaction can be determined and shown in table II

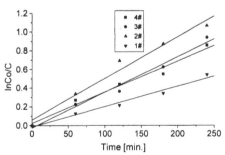

Fig. 5 First-order plot for the kinetics of phenol degradation in the presence of TiO$_2$ films (1$^{\#}$, 2$^{\#}$, 3$^{\#}$, 4$^{\#}$)

Table II The preparation condition of TiO$_2$ films and its photodegradation results on phenol

Sample	Degradation ratio[a] $\Delta C/C_0$	Rate constant (min.$^{-1}$)	Half time $t_{1/2}$ (min.)	Specific photocatalytic acitivity[b] (mol.g^{-1}.h^{-1})
Blank	0.128	5.6 × 10^{-4}	1246	—
1[#]	0.419	2.2 × 10^{-3}	315	1.2 × 10^{-3}
2[#]	0.656	4.6 × 10^{-3}	151	8.5 × 10^{-4}
3[#]	0.610	3.8 × 10^{-3}	182	4.8 × 10^{-4}
4[#]	0.577	3.3 × 10^{-3}	210	1.5 × 10^{-4}

[a] average degradation ratio of phenol after 4 hours photocatalytic reaction

[b] phenol degradation amount per unit weight catalyst after one hour photocatalytic reaction

Table II list the data about photocatalytic activity of TiO$_2$ films. It could be found that the film's activity decreased with its thickness, which was different from the relation shown in Fig.4. This phenomenon can be well explained by the formation process of the film.

As we know, the effective specific surface area was a decisive factor to the activity of a catalyst's. Sample 1 was a coarse particulate film (shown in Fig.1a). Such film had large specific surface area. With the extension of depositing time, the film became more and more dense, more and more smooth, and its specific surface area became more and more small consequently. So, the catalytic activity of sample 1,2,3,4 decreased in turn. However, the practical effect of a catalyst was also related to its amount. Though the sample 1 had the largest specific surface area, it contained the smallest amout of TiO$_2$, and thus influenced its practical degradation rate. So, combined with the two factors, 2[#] had the largest effective surface area in the four samples and so did its degradation ability.

In order to evaluate the activity of the TiO$_2$ films, we also took contrast experiment on P25 (Degussa). The results showed that the activity of sample 1 and 2 was approaching to that of P25 (2.02 × 10^{-3} mol·g^{-1}·h^{-1}). The higher activity may be achieved by optimizing the preparation conditions further.

CONCLUSIONS

By means of PECVD method, a new kind of high-efficient TiO$_2$ film catalyst was prepared. SPS results showed its photoresponse extended to the visible region. Photodegradation experiments on phenol showed the distinct relation between the thickness of film and its catalytic activity. By optimizing preparing conditions, high photodegradation rate can be obtained.

ACKNOWLEDGEMENT

This research was supported by the National Natural Science Fundation of China

REFERENCES

[1] A.L. Linsebigler, G.Q. Lu and J.T. Yates, Jr., Chem. Rev. **95**, p. 735 (1995).

[2] M.R. Hoffmann, S.T. Martin, W.Y. Choi and D.W. Bahnemann, Chem. Rev. **95**, p.69 (1995).

[3] R.W. Matthews, J. Phys. Chem. **91**, p.3,328 (1987).

[4] A.J. Hoffman, G. Mills, H. Yee and M.R. Hoffmannn, J. Phys. Chem. **96**, p.5,546 (1992).

[5] W.Y. Choi, A. Termin and M.R. Hoffmann, J. Phys. Chem. **98**, p.13,669 (1994).

[6] N. Serpone, D. Lawless, J. Disdier and J.-M. Herrmann, Langmuir **10**, p.643 (1994).

[7] K. Vinodgopal, I. Bedja and P.V. Kamat, Chem. Mater. **8**, 2,180 (1996).

[8] J. Mazurowski, S. Lee, G. Ramseyer and P.A. Dowben, Mater. Res. Soc. Symp. Proc. 1992, p.242

[9] D.J. Wang, W. Liu, L.Z. Xiao and T.J. Li, Chemistry (Chinese) **10**, p.32 (1989).

Part II

Metal Catalysts

MOCVD PALLADIUM AND PLATINUM SUPPORTED ON ALUMINA CATALYSTS: PREPARATION AND CHARACTERIZATION

Z.K. LOPEZ*, J.R. VARGAS**, M.A. VALENZUELA*, D.R. ACOSTA***
*Dept. of Chemical Engineering, ESIQIE-IPN, 07738, Mexico, D.F., MEXICO
**Dept. of Metallurgical Engineering, ESIQIE-IPN, AP 75-874, Mexico, D.F., MEXICO
***Instituto de Física. UNAM., A.P. 20-364 Mexico, D.F:, 01000, MEXICO

ABSTRACT

An horizontal hot-wall MOCVD reactor was used to prepare palladium and platinum catalysts supported on alumina. A conventional impregnated Pt on alumina catalyst was prepared as comparison. The solids were characterized by XRD, Auger spectroscopy, HREM and H_2-TPR. The operation conditions of the MOCVD reactor were fixed preparing several Pd catalysts until to find the appropriate deposition zone. The particle size of Pt catalysts prepared by MOCVD was at about 7 nm compared with 6 nm obtained with the Pt impregnated catalyst, measured by XRD. The HREM image of the Pt MOCVD catalyst showed a narrower particle size ranging from 1 to 4 nm. After calcination three Pt compounds were detected by TPR, which were attributed to PtO, PtO_2 and $Pt-Al_2O_3$ interaction in MOCVD preparation. Additionally, a clear reduction of surface oxygens of alumina was also observed.

Keywords: MOCVD, Pt catalysts on alumina, XRD, Auger spectroscopy, HREM, TPR.

INTRODUCTION

Metallic supported catalysts are widely used in dehydrogenation, reforming and other industrial reactions [1]. They are usually prepared by conventional methods such as impregnation, coprecipitation and ion interchange [2]. The main disadvantage using these methods is that after thermal treatments a broad particle size distribution is obtained. In the last decade, efforts have been made in order to prepare catalysts with more homogeneity to improve their catalytic properties. Decomposition of a metal cluster compound, chemical deposition, ion implantation, vapor-phase deposition and sol-gel routes have been reported as novel preparation methods for supported metallic particles [2,3]. Platinum and palladium supported on zeolite catalysts have been already prepared by a modified metal organic chemical vapor-phase deposition combined with a separate step of decomposition under a reduction atmosphere. A small size and thermal stability of metallic particles with the desired topology were obtained by this two-step technique [4]. Decomposition of Pt-acetylacetonate on alumina and on silica in a fluidized-bed reactor was studied in order to correlate the effect of the surface properties and dispersion of metallic particles. The deposited particles were more mobile on silica and more readily agglomerate; on the other hand, dispersion of platinum particles on alumina did not change at high metal loading. After the final thermal treatment the metallic particles were contaminated with carbon [5].

In the present work, we report the structure and catalytic performance of platinum and palladium supported on alumina catalysts prepared by MOCVD. A conventional impregnated platinum supported catalyst was prepared and analyzed for comparison.

EXPERIMENT

Platinum (Pt) and palladium (Pd) catalysts were prepared in a horizontal hot-wall MOCVD apparatus which is depicted in Fig. 1. A commercial γ-Al$_2$O$_3$ powder (Haldor-Topsoe), 80-100 mesh, was used as a support. It was first thermally pretreated to remove water at 400 °C in an argon flow. The powdered support was held by a stainless steel mesh. Metal-acetylacetonates [(CH$_3$-COCHCO-CH$_3$)$_2$Pt, Aldrich 97% and (CH$_3$-COCHCO-CH$_3$)$_2$Pd, Aldrich 99%] were used as precursors. These precursors were kept at constant temperature (T$_{prec}$) of 180 °C. Their vapors were carried by argon gas with a flow rate (FR$_{Ar}$) of 180 ml/min. Oxygen gas (FR$_{O2}$=30 ml/min.) was occasionally added through a separate quartz tube nozzle to avoid carbon co-deposition. Deposition temperatures (T$_{dep}$) were changed between 300 and 600 °C and total gas pressures (P$_{tot}$) were controlled from 0.1 to 10 Torr. Depositions were usually performed for 20 min. The impregnated platinum catalyst was prepared using a H$_2$PtCl$_6$ alcoholic solution to obtain a 3% Pt loading. The excess of solvent was eliminated in a rotatory evaporator at 80 °C. Calcination was carried out in a static atmosphere from 25 to 500 °C at a heating rate of 10 °C/min. The structure of the metal supported catalysts was investigated by X-ray diffraction (XRD) and high resolution electron microscopy (HREM). The surface composition was analyzed by Auger spectroscopy. The H$_2$ reduction behavior was evaluated by a conventional temperature-programmed technique (TPR).

RESULTS

The appropriate deposition conditions in the MOCVD reactor, to obtain a good appearance of the metallic deposits on the surface of the support were determined by preparing several Pd catalysts. Deposition conditions were found as follows: T$_{dep}$=400 °C, P$_{tot}$=1 Torr.

Figure 2 shows the XRD patterns of Pt supported on alumina catalysts. The reflections from γ-Al$_2$O$_3$ are shown as a reference. The catalysts prepared by MOCVD exhibit reflections at 2θ=39.7°, 46.2°, 67.5° and 81.3° along with reflections of γ-Al$_2$O$_3$. This indicates the

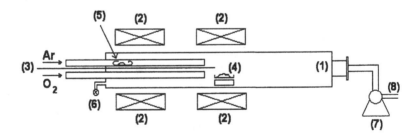

Figure 1. Schematic diagram of the horizontal hot-wall type apparatus.
(1) quartz tube, (2) electric furnace, (3) thermocouple, (4) powdered support,
(5) precursor, (6) manometer, (7) vacuum pump, (8) gas outlet.

presence of the fcc crystal structure of Pt. MOCVD catalysts prepared with oxygen addition (Pt/γ-Al$_2$O$_3$) and without addition (Pt-C/γ-Al$_2$O$_3$) do not show appreciable difference in their XRD patterns. However, Pt-C/γ-Al$_2$O$_3$ had a darker appareance which was associated with a carbon content as revealed by Auger analysis. The impregnated Pt catalyst (calcined at 500°C) showed less intense Pt reflections compared with those of MOCVD catalysts. Subsequent reduction with hydrogen increased the intensity of Pt reflections in the impregnated catalyst.

Irrespective of the preparation method, supported catalysts exhibit broad Pt reflections which suggest a small size of Pt crystallites. The average size of Pt crystallites was estimated from the reflection at $2\theta=39.7°$ by the Scherrer equation. It was about 7 nm for MOCVD catalysts and 6 nm for impregnated catalyst (calcined sample).

The Auger spectra of γ-Al$_2$O$_3$ and MOCVD Pt/γ-Al$_2$O$_3$ are shown in Fig. 3. Only Pt signals were observed on the surface of the MOCVD catalysts prepared with oxygen addition. This indicated that the addition of oxygen was effective to obtain uniform surface coverage of Pt carbon-free deposits. Nevertheless, observable signals of carbon and oxygen along with those of Pt in catalysts prepared without oxygen addition suggested the formation of a metal-carbon composite (Pt-C) on the surface of the support. Ir-C nanocomposites have been previously prepared by MOCVD [6].

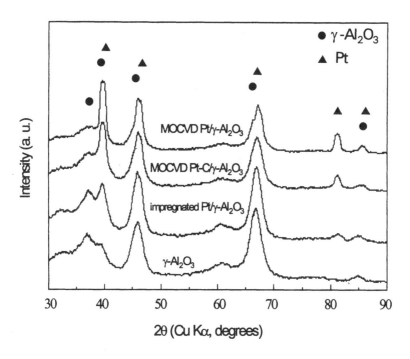

Figure 2. XRD patterns of γ-Al$_2$O$_3$, impregnated Pt catalyst and MOCVD Pt catalysts.

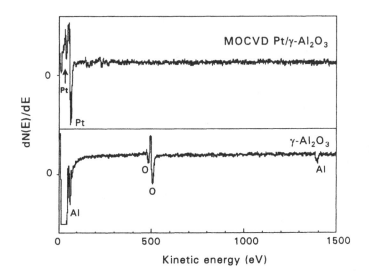

Figure 3. Auger spectra of γ-Al$_2$O$_3$ and MOCVD Pt/γ-Al$_2$O$_3$ catalyst.

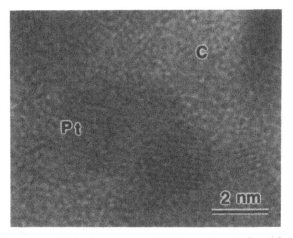

Figure 4. HREM micrograph of MOCVD Pt-C/γ-Al$_2$O$_3$ catalyst.

The distribution of Pt crystallites size on γ-Al$_2$O$_3$ was investigated by electron microscopy. Figure 4 reveals the HREM image of a typical Pt-C/γ-Al$_2$O$_3$ catalyst. Small crystalline Pt particles (1-4 nm) are embedded by an amorphous phase which was associated with the carbon content in these catalysts. MOCVD Pt-C/γ-Al$_2$O$_3$ catalysts actually consist of agglomerated fine Pt particles with a film-like-appearence. The higher Pt particle size estimated by Scherrer equation (XRD) was attributed to the inherent average size determined by this method.

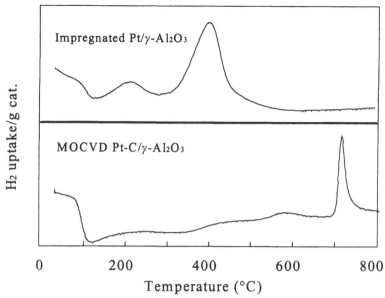

Figure 5. TPR profiles of impregnated Pt catalyst and MOCVD Pt catalyst.

Figure 5 reports the TPR of the MOCVD Pt catalyst calcined in situ at 800°C. The MOCVD sample showed three reduction bands at about 200, 450 and 580°C and one defined peak at 730°C. The first and second bands are characteristic reduction zones of the Pt oxides [7]. The third band could be attributed to a metal-support interaction. The peak observed at high temperature is assigned to the reduction of the surface oxygens of the alumina. The TPR profile of the Pt impregnated catalyst (calcined sample) is also shown in Fig.5. As can be seen, only two well-defined peaks at 220° and 400°C corresponding to PtO and PtO_2 are observed [8] and the net H_2 uptake is closely to the stoichiometric required. In summary, in the MOCVD Pt catalyst the Pt reduction degree is lower compared with the impregnated sample. The $Pt-Al_2O_3$ interaction as well as the surface attack of the support is not observed in the impregnated catalyst.

CONCLUSIONS

The Pd and Pt supported on alumina catalysts were prepared by the MOCVD method. The appropriate deposition conditions were found at T=400°C and P= 1 Torr by preparing Pd catalysts. The effect of oxygen addition in the MOCVD reactor generates a uniform surface coverage of platinum on alumina, whereas no oxygen addition causes a film of agglomerated small Pt particles embedded in amorphous carbon on alumina. Pt particle size obtained by MOCVD was at about 1-4 nm measured by HREM. In spite of the small Pt particles in the MOCVD catalysts, the poorly defined peaks in their TPR profiles suggest that oxidation and reduction processes are difficult due to the complexity of the surface. The catalytic test of these solids will clarify the structure and the performance of Pt particles.

ACKNOWLEDGMENTS

This research was supported by CONACYT (projects 2584-PA and 1988-PA), DEPI-IPN (projects 953466 and 962686) and FIES-IMP-95-34-III.

REFERENCES

1. J.N.Armor, Appl.Catal. **78,** 141 (1991).
2. M.Che and C.O.Bennett, Adv.Catal. **36,** 55 (1989).
3. T.López, P.Bosch, M.Morán, R.Gómez, J.Phys.Chem., **97,** 1671 (1993).
4. C. Dossi, R. Psaro, A. Bortsch, E. Brivio, A. Galasco, P. Losi, Catal. Today, **17,** 527 (1993).
5. S. Köhler, M. Reiche, C. Frobel M. Baerns in <u>Preparation of Catalysts VI. Scientific Bases for the Preparation of Heterogeneous Catalysts</u>, edited by G. Poncelet et al. (1995 Elsevier Science B.V.), p. 1009-1016.
6. T.Goto, J.R. Vargas, T.Hirai, Mat.Sci.Eng., **A217/218,** 223 (1996).
7. G.Aguilar-Ríos, M.A.Valenzuela, H.Armendariz, P.Salas, J.M.Dominguez, D.R.Acosta, I.Schifter, Appl.Catal. A, **90,** 25 (1992).
8. H.Lieske, G.Lietz, H.Spindler, J.Völter, J.Catal. **81,** 8 (1983).

Mechanically Alloyed Nickel-Zirconium as a Heterogeneous Catalyst and a Catalyst Precursor

W. E. Brower, Jr., A. J. Montes, and K. A. Prudlow
Department of Mechanical and Industrial Engineering
Marquette University, Milwaukee, Wisconsin 53233 USA

H. Bakker, A. C. Moleman, and H. Yang
Van der Waals-Zeeman Laboratorium
University of Amsterdam, Valckenierstraat 65, NL-1018 XE
Amsterdam, The Netherlands

ABSTRACT

NiZr powders produced by mechanical alloying become active NO decomposition catalysts after an activation period in reaction conditions. Although the initially glassy structure exhibits a high activity, the NiZr powder becomes temporarily inactive. After about 10 hours in reaction conditions at 673 K, the powder again becomes active with a lower, but still substantial turnover frequency for NO decomposition. Oxygen produced by the reaction appears to be consumed by the catalyst during this second period of activity.

INTRODUCTION

Several excellent reviews of the application of metallic glasses as heterogeneous catalysts have appeared [1-6] as early as 1982 by Giessen et al. [2] and as late as 1994 by Baiker [6]. Glassy metal alloy catalysts have been evaluated in a wide range of reactions, mostly hydrogenation, but also ammonia synthesis, oxidation of CO, dehydrogenation, hydrogenolysis, and NO reduction. The direct decomposition of NO has been studied in this work. Rapidly quenched metal-metalloid alloy glasses were reported to be active catalysts in the early 1980's [2,7,8,9]. The glassy phase exhibited novel selectivity as compared to its crystalline counterpart [8,9]. However, Giessen and coworkers found no such difference [2].

Several methods have been employed to produce metallic glass alloys including rapid quenching from a liquid, solid state interdiffusion of thin layers, mechanical alloying, and precipitation from aqueous salt solutions. Successful hydrogenation catalysts have been produced from precipitated amorphous NiB [10,11]. Mechanically alloyed glasses have been studied by several groups in the last several years as various forms of catalysts. Hout et al. [12] produced microcrystalline NiMo alloys by mechanical alloying which were effective electrocatalysts for hydrogen evolution from KOH solutions. Alves et al. [13] observed mechanically alloyed NiMo to perform as an electrocatalyst as well as platinum. Mulas et al. [14] using NiZr and Trovarelli [15] using CeO_2 and ZrO_2 both produced active catalysts by milling in the presence of carbon monoxide and hydrogen. Streletskii et al. [16] using Zr, NiZr, and Ni, produced active methanation catalysts during milling. Brower et al. [17] reported an active NiZr catalyst for the decomposition of NO, which was formed by mechanical alloying and in vacuum. The turn over frequency was higher on the MA NiZr powder than that on supported Ni and pure Ni powder.

The stability of the surface of a metallic glass is low toward crystallization and oxidation [18,19]. Moreover, the presence of the reactant gases lowers the stability of the surface of metallic glasses still further [6,20,21]. Thus, more recent studies have focused on what the metallic glasses become during reaction conditions. This could be analogous to the "activation" treatments for conventional catalysts. The insightful experiments of the Japanese workers in 1981 [22] led the way to transforming the metallic glasses to active catalysts by a long exposure to reaction conditions. We report here a similar approach, where we have observed the activity of mechanically alloyed NiZr metallic glass as a catalyst for the decomposition of nitric oxide.

EXPERIMENTAL PROCEDURE

NiZr alloys were milled in a vibratory ball mill in 1×10^{-6} vacuum as described by Bakker et al. [22]. Surface area and chemisorption measurements were performed on the Quantasorb Chembet 3000. Surface area was measured by physisorption of nitrogen by the BET method. The number of surface metal atoms per gram was determined by selective chemisorption of carbon monoxide at room temperature after cleaning the surface in flowing hydrogen for 2 hours at 673 K. Decomposition kinetics of nitric oxide flowing at 5 cc/min. were measured in a flow reactor containing mechanically alloyed NiZr powder. The decomposition kinetics were studied under both continuous heating and isothermal conditions. For continuous heating experiments, the reactor was heated at a rate of 10 °C/min with a constant flow rate of 5 cc/min of NO in 30 cc/min of He. Isothermal experiments were performed at 673 K with a constant flow rate of 5 cc/min of NO in 18 cc/min of He. Reactants and products were measured in a Gow Mac gas chromatograph equipped with a thermal conductivity detector and a 6 ft x 1/8 inch Hayesep Db column.

RESULTS and DISCUSSION

Figure 1 shows SEM images of the as milled NiZr powder. The powder is amorphous, as indicated by XRD [17], and exhibits a surface area of 0.1 m^2/g. The surface gives the appearance of severe cold work, characteristic of the milling process, the mechanism for alloy formation at such low temperatures. A rather cracked, porous surface is consistent with the surface area's being several times higher than rapidly quenched metallic glasses. Figure 2 shows the results of NO conversion over NiZr mechanically alloyed powder for continuous heating at 10 °C/min at a constant flow rate of 5 cc/min of NO in 30 cc/min of He. The temperature for 50% conversion, the light off temperature shifts from 292 °C to 393 °C upon cooling and reheating the catalyst. Other examples of this same behavior are shown in Table 1 for $Ni_{50}Zr_{50}$ and $Ni_{45}Zr_{55}$ alloys.

It appears from the decreased activity of the catalysts, as indicated by the increased light off temperatures, that the catalyst is changing and deactivating. The times above light off temperatures for these continuous heating runs were about an hour for each heating run. The initially glassy as-milled structure could be crystallizing during this period. Previous XRD results showed crystallization to a crystalline metastable NiZr phase at 873 K in vacuum and the formation of Ni and ZrO_2 in air at 1073 and 673 K, respectively [17]. To investigate the change in the metastable milled powder, an isothermal reactor run was carried out at 673 K.

Figure 3 shows the results of this isothermal conversion of NO over a $Ni_{55}Zr_{45}$ mechanically alloyed powder catalyst. After an initial 2-3 hour period of conversion of NO and the production of N_2 but no O_2, the catalyst becomes inactive at the 673 K reaction temperature. The same deactivation appears to be occurring during the continuous heating runs in Table 1, as

Figure 1. SEM micrographs of NiZr as-milled powder after 20 hours milling time. Left - 1600x, Right - 5000x.

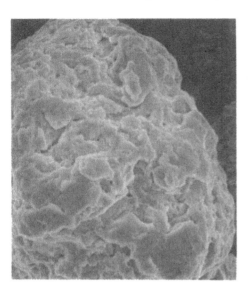

Figure 2. NO Conversion of Reactor 1 - Ni$_{45}$Zr$_{55}$ Mechanically Alloyed Powder.

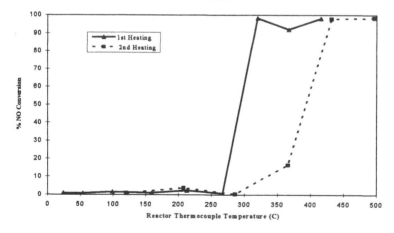

Table 1. Lightoff Temperatures of NiZr Catalyst Reactors.

| | $Ni_{45}Zr_{55}$ Reactor 1 | | $Ni_{45}Zr_{55}$ Reactor 2 | | $Ni_{50}Zr_{50}$ Reactor 3 | | $Ni_{50}Zr_{50}$ Reactor 4 | |
	1st heating	2nd heating	1st heating	2nd heating	1st heating	2nd heating	1st heating	2nd heating
$T_{lightoff}$ (50% Conversion)	292°C	393°C	550°C	700°C	306°C	480°C	395°C	465°C

Figure 3. Activation of Mechanically Alloyed $Ni_{55}Zr_{45}$ Powder for NO Decomposition at 673 K. (Sample Mass = 0.3137 g, NO flow = 5 cc/min., He flow = 18 cc/min.)

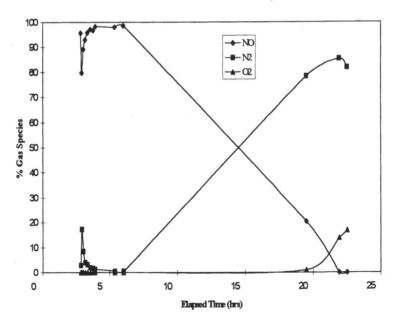

evidenced by the increased light off temperatures upon second heating. The calculated TOF for this transient active period is 15.5/sec, comparable to the TOF after reaction in the continuous heating runs, Table 2, of 12.2/sec. This calculated TOF assumes a surface metal concentration of 9.2×10^{16} atoms/g for the NiZr catalyst of Figure 3 after 2 hrs reaction time, the value given in Table 2 for NiZr "after NO reaction", a time in reaction of about 2 hrs. After about 10 hours, the activity of the catalyst increases as evidenced by the increased conversion of NO, reaching 80%, and the appearance of N_2, followed by the appearance of O_2 for the first time. At about 23 hours of reaction time, the vycor reactor tube fractured, so the run was terminated and the catalyst examined. Table 2 shows the results of chemisorption measurements and turn over frequency (TOF) for the catalyst from Figure 3, along with previous continuous heating results for comparison [17]. Our attempt at measuring the surface area via the BET technique with N_2 as the adsorbate indicated the surface area was too low to be measured. CO chemisorption measurements of the surface metal content yielded a value of 9.45×10^{18} atoms/g. From this number and the rate of conversion of NO, a TOF was calculated to be 0.83/sec at 80% conversion of NO. This second value of TOF for the NiZr powder at 23 hrs reaction time is well above that measured for supported Ni and Ni powder, Table 2. As the reaction proceeds, ZrO_2 appears to be forming from product oxygen, probably leaving behind Ni particles. Thus, a type of supported Ni/ZrO_2 catalyst forms, the Ni forming in the former NiZr matrix, not precipitating on the support, as is the case with the conventional supported Ni catalyst. The higher TOF of the mechanically alloyed NiZr powder may be due to an enhanced "SMSI", strong metal-support interaction, as a result of ZrO_2 and Ni forming from the as milled glassy NiZr precursor solid solution.

Table 2. Surface Area, Chemisorption, and Turnover Frequency Results for Catalysts Studied for NO Reaction.

Alloy	Surface Area (m^2/g)	# of Surface Metal Atoms/g	Temperature (K)	Turnover Frequency $(sec.^{-1})$
Supported Ni-SiO_2/Al_2O_3*	189	1.56×10^{16}	473	7.2×10^{-4}
Ni Powder*	2.28	2.28×10^{16}	803	0.18
$Ni_{50}Zr_{50}$ (as-received)*	0.11	3.81×10^{16}	573	29.4
$Ni_{50}Zr_{50}$ (after NO reaction)*	3.2	9.20×10^{16}	573	12.2
$Ni_{55}Zr_{45}$ (after NO reaction)	off scale	9.45×10^{18}	673	0.83

* Reference 17

CONCLUSIONS

Yamashita and Vannice [23] reported on NO decomposition kinetics over an MnO catalyst. No activity was reported below 773 K, whereas the NiZr work we report here shows conversion at 673 K. An activation period appears to convert the NiZr to supported Ni on zirconia.

ACKNOWLEDGMENTS
The authors would like to thank Briggs and Stratton Corporation for partial support of this work and the Netherlands Stichting FOM for support of one of the authors (WEB) while on sabbatical.

REFERENCES

1. A. Baiker, Faraday Discuss. Chem. Soc. 87, 239 (1989).
2. B.C. Giessen, S. S. Mahmoud, D.A. Forsyth, and M. Hediger in Rapidly Solidified Amorphous and Crystalline Alloys edited by B. H. Gear, B. C. Giessen, and M. Cohen (Elsivier Science Publishing Co., Inc., New York, 1982), pp. 255-258.
3. D. L. Cocke, J. Metals, Feb., 71 (1986).
4. R. Schloegl in Rapidly Quenched Metals edited by S. Steeb and H. Warlimont, Elsevier Science Publishers, B. V. (1985), pp. 1723-1727.
5. C. Yoon and D. L. Cooke, in Rapidly Quenched Metals edited by S. Steeb and H. Warlimont, Elsevier Science Publishers, B. V. (1985), pp 1497-1504.
6. A. Baiker in Topics in Applied Physics, Vol. 72, Beck and Guntherodt editors (1994), pp. 121-162.
7. A. Yokoyama, H. Komiyama, H. Inoue, T. Matsumoto, and H. M. Kimura, J. Catal. 68, 335 (1981).
8. G. V. Smith, W. E. Brower, Jr., M. Matyjasczcyk, and T. L. Pettit, Proc. of 7th Int. Cong. on Catalysis, Tokyo, 1980.
9. R. Hauert, P. Oelhafen, R. Schlogl and H. J. Guntherodt in Rapidly Quenched Metals Edited by S. Steeb and H. Warlimont (Elsevier Science Publishing Co., B. V., 1985) p. 1493.
10. G. Carturan, S. Enzo, R. Canzeria, M. Lenarda, and R. Zanoni, J. Chem. Soc. Faraday Trans. 86, 739 (1990).
11. J. Deng, J. Yang, S. Sheng, H. Chen, and G. Xiong, J. Catal. 150, 434 (1994).
12. J. Y. Hout, M. L. Trudeau, and R. Schultz, J. Electrochem. Soc. 138, 1316 (1991).
13. H. Alves, M. Ferreira, and U. Koster, Materials Science forum 179-181, 494 (1995).
14. G. Mulas, L. Conti, G. Scano, L. Schiffini, and G. Cocco, Mats. Sci and Engrg. A181/A182, 1085 (1994).
15. A. Trovarelli, P. Matteazzi, G. Dolcetti, A. Lutman, and F. Miana, Applied Catal. 95, 19 (1993).
16. A. N. Streletskii, O. S. Morozova, I. V. Beresteskaja, and A. B. Borunova, Proc. of Int. Symp. on Metastable, Mechanically Alloyed and Nanocrystalline Materials, Barcleona, 1997.
17. W. E. Brower, Jr., A. J. Montes, K. A. Prudlow, H. Bakker, A. C. Moleman, and H. Yang, Proc. of Int. Symp. on Metastable, Mechanically Alloyed and Nanpcyrstalline Materials, Rome, 1996.
18. F. Buffa, A. Corrias, G. Licheri, G. Navarra, and D. Raoux, J. Noncrystalline Solids 150, 386 (1992).
19. H. J. Guntherodt in Rapidly Quenched Metals edited by S. Steeb and H. Warlimont Elsevier, New York, 1985, p. 1591.
20. W. Kowbel and W. E. Brower, Jr., J. Catal. 101, 262, (1986).
21. T. Yamashita and A. Vannice, J. Catal. 153, 158 (1996).
22. H. Bakker, G. F. Zhou, and H. Yang, Prog. in Materials Science 39, 159 (1995).

CHEMICAL CONTROL OF NOBLE METAL CATALYSIS BY MAIN GROUP ELEMENTS

K. Asakura*, K.Okumura**, T.Inoue**, T.Kubota**, W-J.Chun** and Y. Iwasawa**
*Research Center for Spectrochemistry, Faculty of Science, the University of Tokyo, Hongo, Bunkyo-ku, Tokyo 113-0033, Japan, askr@chem.s.u-tokyo.ac.jp
**Department of Chemistry, Graduate School of Science, The University of Tokyo, Hongo, Bunkyo-ku, Tokyo 113-0033, Japan, iwasawa@chem.s.u-tokyo.ac.jp

ABSTRACT

The catalytic interaction of noble metal and main group elements in Rh/one-atomic layer GeO_2/SiO_2 and Pt/SbO_x was investigated. The high temperature reduction produced RhGe and PtSb bimetallic particles in which Pt and Rh were electronically modified to retard catalytic activity. However, unique selective catalyses of Rh/one-atomic layer GeO_2/SiO_2 for CO hydrogenation reaction to oxygenate compounds and for NO+CO reaction to N_2 were found. Under the low temperature reduction of Rh/one-atomic layer GeO_2/SiO_2 and the high temperature calcination of Pt/SbO_x, the oxide phases, GeO_2 and SbO_x, were stable and the selective reduction of ethylacetate to ethanol and the selective oxidation of iso-C_4H_{10} to methacrolein were observed. The high selectivities were ascribed to synergistic interaction between the noble metals and the main group element oxides through the diffusion of adsorbed species and reaction intermediates. The possibility of chemical control of noble metal-catalyses by main group elements is discussed.

INTRODUCTION

Main group elements have often been used as additives for the modification of catalysis of noble metals. Heavy 13-16 main group elements such as Se and Sn show unique synergistic catalyses when they are present on the surface of noble metal particles. For example, Se^0 adsorbed on Rh particles promoted CO insertion reaction[1,2] and Sn^0 on Pt and Rh particles enhanced the NO decomposition reaction[3-5]. Rh-Ge,-Sn, and -Pb were also active for the selective ester and citral hydrogenation to alcohol[6,7]. In these catalysts, the main group elements in low valence states have been demonstrated to have direct bondings with noble metal atoms to modify the active metal sites electronically and geometrically as well as to provide new active sites owing to their oxophillic properties[1, 4, 5]. On the other hand, oxides of heavy 13-16 main group elements have accepted interests as one component of mixed oxides catalysts such as MoBi, [8] FeSb, VSb, [9-12], SnSb [13, 14] oxides which are used for selective oxidation. In these catalysts, the main group elements work as oxygen donor sites, hydrogen extraction sites and adsorbed sites of allyl species[8, 9, 13, 14]. The reaction sites may be coupled with each other through the diffusion of adsorbed species and reaction intermediates. From these points of view, it is possible to prepare active surfaces composed of noble metal particles and main group metal oxides for selective oxidation and hydrogenation, where the reducibility of heavy 13-16 main group metal oxides may regulate the catalysis of noble metal particles. In this paper we will report our recent investigation of synergistic modifications of Rh and Pt catalyses by one-atomic layer GeO_2 and bulk SbO_x respectively[15-22].

EXPERIMENTAL

Material and preparation

The one-atomic layer GeO_2 on SiO_2 (aerosil 300 pretreated at 473 K for 1 h) was prepared by the reaction of surface OH groups of SiO_2 with $Ge(OMe)_4$[16]. The obtained sample was calcined at 693 K for 1 h under 20.0 kPa of oxygen in a closed circulating system. The loading of Ge was 7.4 wt% which was a maximum loading by one cycle deposition of $Ge(OMe)_4$. Then the one-atomic layer GeO_2 on SiO_2 was impregnated with a $Rh_6(CO)_{16}$ chloroform solution under Ar atmosphere[19]. Rh loading was 2.0 wt%. The sample thus obtained was reduced at given temperatures for 2 h under 13.3 kPa of hydrogen.

The SbO_x support was prepared by hydrolysis of $SbCl_5$ with aqueous ammonia solution, followed by calcination at 773 K. Pt was supported on the SbO_x by an impregnation method using an acetone solution of $Pt(acac)_2$, followed by drying at 323 K and calcination at 773 K. Pt loading was 0.5 wt%[17, 22].

Reaction

CO hydrogenation was carried out at $P_{CO}=P_{H2}=13.3$ kPa and 523 K in a closed circulating system. NO reduction by CO was carried out at $P_{NO}=P_{CO}=4$ kPa and 423 K in a closed circulating system. The 0.3 g Rh/one-atomic layer GeO_2/SiO_2 was used for both reactions. The ethylacetate hydrogenation reaction on the 0.1 g Rh/one-atomic layer GeO_2/SiO_2 catalyst was conducted with a mixture of 6.6 kPa of hydrogen and 1.3 kPa of ethylacetate at 473 K in a closed circulating system. Steady state activities for partial oxidation reactions of iso-C_4H_{10} and i-C_4H_8 were examined in a flow line under the conditions of i-C_4H_{10} 20 %, O_2 4 % and 2400 ml h^{-1} for i-C_4H_{10} oxidation and under the conditions of i-C_4H_8 1.7 %, O_2 4% and 2400 ml h^{-1} for i-C_4H_8 oxidation. The 0.3 g Pt/SbO_x catalyst was used for the selective oxidation reactions.

RESULTS

Rh/GeO_2
Structure of Rh/one-atomic layer GeO_2/SiO_2
Rh/bulk GeO_2 was easily reduced at 473 K to form Ge^0 bulk into which Rh was dissolved and did not show any catalytic activity for ethylacetate hydrogenation. In our previous studies, one-atomic layer Nb, Ti, and Zr oxides were hardly reduced even in the presence of Pt and Pd by H_2 reduction at 773 K[23-25]. We tried to stabilize the GeO_2 phase by supporting one-atomic layer GeO_2 structure on SiO_2 surface, using the reaction of $Ge(OMe)_4$ and OH groups of SiO_2. The 7.4 wt % one-

atomic layer GeO₂/SiO₂ showed no XRD peak except a broad one around $2\theta = 21°$ due to amorphous SiO_2. EXAFS analyses of the one-atomic layer GeO₂/SiO₂ revealed the Ge-O and Ge-Ge distances at 0.173 nm and 0.311 nm, respectively. Rh was then deposited on the one-atomic layer GeO₂/SiO₂ from $Rh_6(CO)_{16}$. Transmission electron microscopy indicated that Rh average particle sizes in the Rh/one-atomic layer GeO₂/SiO₂ catalysts reduced at 523 k and 723 K were 2.6 and 2.8 nm, respectively. Fig.1 shows the Fourier transform of Rh K-edge EXAFS spectra for Rh/one-atomic layer GeO₂/SiO₂. The Fourier transform of the sample reduced at 523 K gave a peak corresponding to metallic Rh-Rh bonding. The Rh-Rh distance was determined at 0.266 nm by curve fitting analysis, indicating the formation of Rh particles. When it was reduced at 723 K, the peak shifted to lower distance and the peak

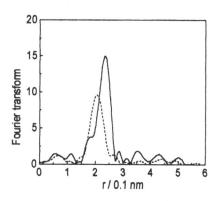

Fig.1 Fourier transformation of Rh K-edge k^3-weighted EXAFS oscillations for Rh/one-atomic layer GeO₂/SiO₂ reduced at 523 K(——) and 723 K(-----).

height decreased. Curve fitting analysis revealed that the peak was not due to the Rh-Rh but Rh-Ge at 0244 nm. We could not find Rh-Rh contribution. The Ge K-edge EXAFS analysis of the Rh/one-atomic layer GeO₂/SiO₂ reduced at 523 K showed that the one-atomic layer GeO₂/SiO₂ was maintained by the reduction at 523 K[19]. The sample reduced at 723 K, the one-atomic layer structure was mainly retained but the small contribution of Ge-Rh bonding was observed at 0.243 nm[19]. Thus the reduction at the lower temperatures < 523 K formed metallic Rh particles on the one-atomic layer GeO_2, while the treatment at the higher temperatures produced GeRh alloy particles where Ge was supplied from one-atomic layer GeO_2 on SiO_2.

Reaction of the Rh/one-atomic layer GeO₂/SiO₂
(a) *CO + H₂ Reaction*

Fig.2 shows the catalytic activity and the selectivity to oxygenated compounds on Rh /one-atomic layer GeO₂/SiO₂ for CO hydrogenation. The Rh/one-atomic layer GeO₂/SiO₂ reduced at 523 K mainly produced CH_4 where the selectivity to oxygenated compounds was as low as 29 %. The activity of Rh / SiO₂ showed 3 times higher than the Rh one-atomic layer GeO₂/SiO₂ under the identical reaction condition, but Rh/SiO₂ produced almost no oxygenates. Total activity of CO hydrogenation on the Rh/one-atomic layer GeO₂/SiO₂ reduced at 723 K decreased to 1/25 of that on the Rh/one-atomic layer GeO₂/SiO₂ reduced at 523 K though the selectivity to oxygenated compounds (C_2H_5OH 60 %, CH_3CHO 30 %, CH_3OH 10 % in all oxygenated compounds) remarkably increased to ca 90 %.

(b) NO +CO Reaction

Fig.3 shows the initial rate of NO reduction with CO as a function of the prereduction temperature of the catalyst. The Rh/one-atomic layer GeO_2/SiO_2 catalyst reduced at 423 K produced mainly N_2O similar to the case of Rh/SiO_2. Increasing the prereduction temperature enhanced the activity and the selectivity to N_2 production (from 17 % for 423 K-reduced sample to 59 % for 723 K-reduced one). The activity took maximum at 623 K and then decreased, but the selectivity towards N_2 formation increased with the increase of prereduction temperature.

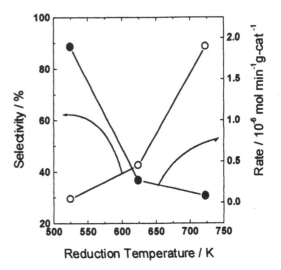

Fig.2 CO hydrogenation activity and selectivity to oxygenates on the Rh/one-atomic layer GeO_2 /SiO_2 at 523 K

(c) Ethylacetate hydrogenation reaction

Fig. 4(a)(b) shows the initial rate of ethylacetate hydrogenation reaction and the selectivity to ethanol formation. The ethanol selectivity is defined as carbon-base selectivity which is given in eq.(1)

$$S_{C_2H_5OH} = \frac{r(C_2H_5OH)}{r(C_2H_5H) + r(CH_3CHO) + 1/2 \cdot r(CH_4) + r(C_2H_6)} \quad (1)$$

On Rh/SiO_2, CH_4 and C_2H_6 were mainly formed as shown in Fig.4(b). The Rh/one-atomic layer GeO_2/SiO_2 produced ethanol with 80 % selectivity when the sample was reduced at 473 K. When the sample was reduced at the higher temperatures acetaldehyde was formed and the ethanol selectivity decreased.

Pt/SbO$_x$

Characterization of Pt/SbO$_x$

X-ray powder diffraction showed the formation of Sb_6O_{13} after calcination of the support SbO_x. When the 0.5wt% Pt/SbO$_x$ was reduced at 473 K, the support was reduced to form Sb_2O_4. EXAFS analysis showed the formation of PtSb alloy particles with the distances of Pt-Sb and Pt-Pt at 0.260 and at 0.274 nm, respectively by the 473 K reduction[17]. On the other hand, when Pt/SbO$_x$ was oxidized at 773 K, only the Pt was reduced to Pt particles with Pt-Pt =0.275 nm and the SbO$_x$ was present in the form of Sb_6O_{13}. Transmission electron microscopy showed that average Pt particle size was 5.5 nm. After iso-C_4H_{10} and iso-C_4H_8 oxidation reactions the Sb_6O_{13} and Pt particles were still observed by XRD and EXAFS. When the catalyst was reacted under more iso-C_4H_8(20%) conditions, the Sb_2O_4 was formed. Table I shows the dispersion of Pt particles in Pt/SbO_x. Assuming a spherical shape of Pt particles, the particle size was estimated to be 13 nm, twice as large as that obtained from transmission electron microscopy. The adsorption of hydrogen was suppressed after the steady state reaction

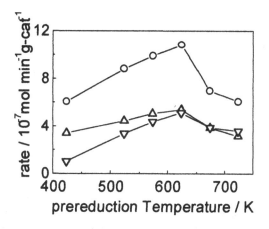

Fig.3 Activity of the Rh/one-atomic layer GeO_2/SiO_2 for NO-CO reaction as a function of prereduction temperature.
$\bigcirc CO_2$:　$\triangle N_2O$:　$\triangledown N_2$

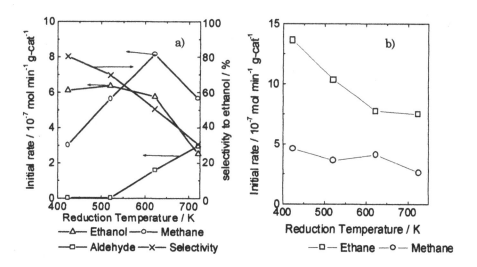

Fig.4 Reduction temperature dependence of the initial rate and selectivity of ethylacetate hydrogenation reaction.

of iso-C_4H_8 and iso-C_4H_{10} and the reduction at 473 K. Raman spectra for the catalysts after the steady state of iso-C_4H_8 oxidation indicated the presence of Sb_6O_{13} (416, 476 and 556 cm^{-1})[20]. Intense peaks were observed at 134, 200, 264, 384, 460, and 718 cm^{-1}, which are assigned to Sb_2O_3. Sb_2O_3 was also visible in the Raman spectra of Pd/SbO$_x$. Another Raman feature characteristic of Pt/SbO$_x$ after the steady state reactions was 500 cm^{-1} and 800 cm^{-1} which are tentatively assigned to Pt-O and to O=SbO$_x$, respectively.

Iso-C_4H_{10} oxidation reaction

Fig. 5 shows the conversion and selectivity to methacrolein in iso-C_4H_{10} and iso-C_4H_8 oxidation reactions on Pt/SbO$_x$ with various reaction temperatures[22]. The reactions were carried out under increasing reaction temperature at a rate of 1.7 K min^{-1}. At low temperatures, total oxidation occurred preferentially. The conversion decreased at 650 K and the methacrolein formation started at 700 K. At 773 K the selectivity to methacrolein reached 50 %. Table II shows the product distribution and selectivity under the steady state reaction conditions at 773 K. Methacrolein and iso-C_4H_8 were formed with 56.6 % and 24.9 % selectivities on 0.5 wt % Pt/SbO$_x$, respectively. On the other hand SbO$_x$ itself showed no activity for iso-C_4H_{10} oxidation. Total oxidation was observed for Pd/SbO$_x$, Rh/SbO$_x$, Ir/SbO$_x$, and Ag/SbO$_x$ under the same conditions. Thus the selective methacrolein formation is due to the synergistic effect of Pt and SbO$_x$. Fig. 5b shows the temperature programmed oxidation reaction of iso-C_4H_8. Iso-C_4H_8 was oxidized to methacrolein with 90 % selectivity on SbO$_x$ surface at 700 K. Pt/SbO$_x$ showed the similar selectivity but the temperature necessary for methacrolein formation became by 50 K lower than that on SbO$_x$. The Pt/SbO$_x$ catalyst reduced with H_2 at 473 K showed little activity for total oxidation and methacrolein formation in iso-C_4H_{10} oxidation.

Table I. Dispersion(H/Pt) for Pt/SbO$_x$ catalysts after calcination, H_2 reduction, and catalytic reactions.

Pt/SbO$_x$	After calcination at 773 K	H_2 reduction at 473 K	iso-C_4H_8 reaction	iso-C_4H_{10} reaction
0.5 wt %	0.07	0.00	0.00	0.00
2.0 wt %	0.24	0.00	0.01	0.01

T_{ads}=273 K, P_{H2}= 13.3 kPa

Table II. Product yields and selectivities for the iso-C_4H_{10} oxidation reaction on the Pt/SbO$_x$ catalysts.

Catalyst	Yield / %				Selectivity / %	
	methacrolein	iso-C_4H_8	CO_2	C_3	methacrolein	iso-C_4H_8
Pt/SbO$_x$ 0.2 wt%	0.15	0.15	0.0	0.04	43.8	43.3
	0.98	0.43	0.24	0.08	56.6	24.9

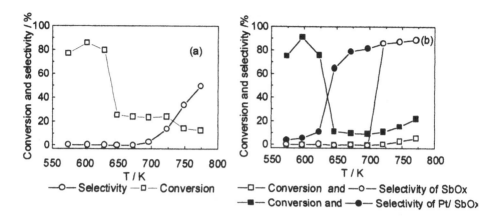

Fig. 5 The temperature programmed reaction profiles on Pt/SbO$_x$ for iso-C$_4$H$_{10}$ oxidation(a) and on Pt/SbO$_x$ and SbO$_x$ for iso-C$_4$H$_8$ oxidation(b); catalyst =0.3 g, total flow rate =2400 ml h-1, iso-C$_4$H$_{10}$ =1.7 %, iso-C$_4$H$_8$=20 %, O$_2$:4 % balanced with He. The heating rate was 1.7 K min^{-1}.

DISCUSSION

Alloying

The structure of main group metal oxides can be controlled by choosing the preparation conditions and pretreatment conditions. GeO$_2$ was stabilized by preparing one-atomic layer structure on SiO$_2$. The oxidation pretreatment and in-situ treatment under the reaction conditions created the selective Pt/SbO$_x$ catalysts for methacrolein production in iso-C$_4$H$_{10}$ and iso-C$_4$H$_8$ oxidation. RhGe alloy particles were formed by the reduction of Rh/one-atomic layer GeO$_2$/SiO$_2$ at 723 K. In the Pt/SbO$_x$, Pt-Sb bonding was easily formed after the H$_2$ reduction at 473 K. The Rh-Ge and Pt-Sb distances were 0.244 nm and 0.260 nm which are much shorter than the values estimated from the sum of the radii of each element. This is due to the difference in electronegativity between the noble metal and the main group atoms, and strong chemical bonds between then are formed. Judging from the electronegativities, electron is transferred from the noble metal to main group atoms, in the same direction as that for well known poisons, S and halogens. Thus the activity of the noble metals is found to be suppressed for CO hydrogenation and NO+CO reaction on Rh/one-atomic GeO$_2$/SiO$_2$ reduced at 723 K and total oxidation reaction on Pt/SbO$_x$ reduced at 473 K. Another feature accompanied with alloy formation is isolation of noble metal ensembles and creation of new ensembles composed of noble metal and main group atoms. In the Rh/one-atomic layer GeO$_2$/SiO$_2$ reduced at 723 K, we found only Rh-Ge bonding, indicating that the Rh atoms were isolated by surrounding Ge. The same isolation effect of noble metal were observed for Pt-Sn/SiO$_2$ catalyst prepared from CVD of Sn(CH$_3$)$_4$ on Pt/SiO$_2$[5], where Pt dimers on the surface of particles were surrounded by Sn atoms. It was suggested that the allyl species in NO-C_3H_6 reaction to produce C_2H_3CN was stabilized on the Sn sites of $PtSn/SiO_2$. Sn species also promoted the dissociation of NO to N_2. Similar effect of Ge was observed in the NO+CO reaction on the RhGe system reduced at high temperatures.

Synergistic effect of main group elements and noble metal

In the Rh/one-atomic GeO_2 /SiO_2 catalyst reduced at low temperatures and the Pt/SbO_x catalyst calcined at 773 K, the chemical bonding between the noble metal and the main group element was not obvious. However, the catalyses were modified by the main group oxides. Examples can be found in CO hydrogenation and ethylacetate hydrogenation on Rh/one-atomic layer GeO_2/SiO_2 reduced and in iso-C_4H_{10} and iso-C_4H_8 selective oxidation reactions on Pt/SbO_x.

One explanation of catalysis modificaiton is the electronic weak interaction between main group atoms and noble metal through charge transfer or the formation of (noble metal) - O - (main group atom) bonding. But details of such interaction are not clear in this work.

Another modification is SMSI like interaction where partially reduced oxide decorates the particle surface [26,27]. In Pt/SbO_x the adsorption capability for H_2 decreased after the reaction at 773 K which may occur by the physical coverage of Sb species. Such modification of Pt surface by Sb species also caused the suppression of total oxidation reaction. In the Raman spectra, we found the formation of Sb_2O_3 and $O=SbO_x$ species during the course of the iso-C_4H_8 oxidation reaction. The former species was also observed in Pd/SbO_x, which catalyzed only the total oxidation, indicating that Sb_2O_3 was unlikely to be the origin for the selective oxidation catalysis. Thus we tentatively attributed the decrease of total oxidation to the dispersion of $O=SbO_x$ species on the Pt surface.

The other effect of the main group oxide is to provide new active sites which are created by the interaction with noble metal. The Rh/one-atomic layer GeO_2/SiO_2 catalyst produced selectively ethanol from ethylacetate, whereas Rh/SiO_2 and one-atomic layer GeO_2/SiO_2 did not show such activity. Fig. 6 shows FT-IR spectra for the ethylacetate adsorbed on the one-atomic layer GeO_2/SiO_2 and the Rh/one-atomic layer GeO_2/SiO_2. The peak at 1728 cm^{-1} on the one-atomic layer GeO_2 /SiO_2 in Fig.6(a) is attributed to unidentate acetate[28]. The peaks observed at 2980, 2934, and 2901 cm^{-1} are assigned to CH-stretching vibration of ethoxy and acetate species, respectively. Thus ethylacetate is adsorbed dissociatively on the one-atomic GeO_2/SiO_2[19]. However, the one-atomic GeO_2/SiO_2 itself was not active for the hydrogenation because it has no ability to activate hydrogen. Rh particles can dissociatively adsorb hydrogen and supply the spilt-over hydrogen to the ethoxy and acetate species on the one-atomic layer GeO_2. Fig. 6(b) and (c) show that the adsorbed ethoxy and acetate decreased by the H_2 treatment. Such coupling between different sites through the diffusion of adsorbed species was also observed on Pt/SbO_x during the oxidation reaction of iso-C_4H_{10}. Sb_6O_{13} was active for iso-C_4H_8 oxidation to form methacrolein though it was deactivated because the reoxidation of the Sb oxide hardly occurred[9]. The active site of Sb_6O_{13} is lattice oxygen as demonstrated by the pulse experiment of iso-C_4H_8[20]. The spillover oxygen from the Pt particles reproduces the active lattice oxygen of Sb_6O_{13} in Pt/SbO_x. In the Raman spectra, the peak at 500 cm^{-1} attributable to Pt-O was observed for the Sb-modified Pt particles under the reaction conditions. The other function of Pt is the dehydrogenation of iso-C_4H_{10} to produce iso-C_4H_8 and (iso-C_4H_7 allyl species). No activity for the dehydrogenation reaction of iso-C_4H_{10} was observed on SbO_x. The iso-C_4H_8 and iso-C_4H_7 species created on Pt particle is transferred to the Sb_6O_{13} phase and oxidized by the lattice oxygen of Sb_6O_{13} to form methacrolein.

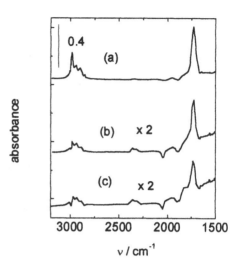

Fig.6 FT-IR spectra of ethylacetate adsorbed on one-atomic layer GeO_2/SiO_2 (a) at 473 K after the exposure of 1.3 kPa of ethylacetate, followed by evacuation of the gas phase at the same temperature. (b) and (c) are FT-IR spectra of $Rh/GeO_2/SiO_2$ exposed to ethylacetate and evacuated and treated with H_2 for 10 min at 473 K.

CONCLUSIONS

Ge and Sb interact with noble metal in various ways, which can be controlled by appropriate chemical procedures. They can be reduced to form strong bondings with Rh and Pt metals. The formation of direct bonds modifies the metal activity and isolates the metal sites. Moreover, the main group elements provide new reaction sites for dissociation of NO. By low temperature reduction, the one-atomic layer GeO_2 on SiO_2 was stable and strongly interacted with Rh particles. The resultant Rh/one-atomic layer GeO_2/SiO_2 catalyst showed the selective hydrogenation property for ethylacetate to ethanol. Further, we found the SMSI-like phenomenon in the Pt/SbO_x catalyst under the oxidative reaction conditions, where the Pt particle surface was covered by $SbO_y(y<x)$ and $O=SbO_x$, resulting in suppression of the total combustion of iso-C_4H_{10} and iso-C_4H_8. The modified Pt surface had the capability enough for the dehydrogenation of iso-C_4H_{10} and iso-C_4H_8. As a result, the Pt/SbO_x catalyst composed of the modified Pt particles and Sb_6O_{13} showed the selective catalysis for the iso-C_4H_{10} and iso-C_4H_8 oxidation reactions. The chemical control of the states and structures of noble metals by main group oxides may provide a new way to regulate the catalytic performance and to prepare new catalytic materials.

ACKNOWLEDGMENT

This work has been supported by CREST(Core Research for Evolutional

Science and Technology) of Japan Science and Technology Corporation(JST).

REFERENCES

1. Y. Izumi, K. Asakura, and Y. Iwasawa, J.Catal., **127**, 631 (1991).
2. Y. Izumi and Y. Iwasawa, J.Phys.Chem., **96**, 10942 (1992).
3. K. Tomishige, K. Asakura, and Y. Iwasawa, J. Chem. Soc., Chem. Commun., 184 (1993).
4. K. Tomishige, K. Asakura, and Y. Iwasawa, Chem.Lett., 235 (1994).
5. T. Inoue, K. Tomishige, and Y. Iwasawa, J. Chem. Soc. Faraday Trans., **92**, 461 (1996).
6. A. E. Mansour, O. A. Ferretti, J. M. Basset, J. P. Bournonville, and J. P. Candy, Angew.Chem. Intern.Ed., **28**, 347 (1989).
7. B. Didillon, J. P. Candy, F. Lepeletier, O. A. Ferretti, and J. M. Basset, Stud. Surf. Sci. Catal., **78**, 147 (1993).
8. Y. Moro-oka and W. Ueda, Adv. Catal., **40**, 233 (1994).
9. G. Centi and F. Trifiro, Catal.Rev.Sci.Eng., **28**, 165 (1986).
10. M. Bowker, C. R. Bicknell, and P. Kerwin, Appl.Catal., **136**, 205 (1996).
11. J. Nilsson, A. Landacanovas, S. Hansen, and A. Andersson, Catal.Today, **33**, 97 (1997).
12. A. Andersson, S. L. T. Andersson, G. Centi, R. K. Grasselli, M. Sanati, and F. Trifiro, in Proc.10th Intern. Congr. Catal.,(Budapest, Hungary, 1992), p 691.
13. L. T. Weng, P. Ruiz, and B. Delmon, in New Developments in Selective Oxidation by Heterogeneous Catalysis, edited by P. Ruiz and E. Delmon,.(Elsevier, Amsterdam, 1992) p. 399.
14. L. T. Weng, N. Spitaels, B. Yasse, J. Ladriere, P. Ruiz, and B. Delmon, J.Catal., **132**, 319 (1991).
15. K. Okumura, K. Asakura, and Y. Iwasawa, J.Mol.Catal., in press.
16. K. Okumura, K. Asakura, and Y. Iwasawa, Langmuir, in press.
17. T. Inoue, K. Asakura, and Y. Iwasawa, J.Catal. in press.
18. K. Okumura, N. Ichikuni, K. Asakura, and Y. Iwasawa, J.Chem.Soc.,Faraday Trans., **93**, 3217 (1997).
19. K. Okumura, K. Asakura, and Y. Iwasawa, J.Phys.Chem. (1997).
20. T. Inoue, K. Asakura, W. Li, S. T. Oyama, and Y. Iwasawa, Appl.Catal., in press.
21. K. Okumura, K. Asakura, and Y. Iwasawa, Chem. Lett., 985 (1997).
22. T. Inoue, K. Asakura, and Y. Iwasawa, J.Catal., **171**, 184 (1997).
23. K. Asakura, M. Aoki, and Y. Iwasawa, Catal.Lett., **1**, 395 (1988).
24. K. Asakura and Y. Iwasawa, J.Phys.Chem., **95**, 1711 (1991).
25. K. Asakura, J. Inukai, and Y. Iwasawa, J.Phys.Chem., **96**, 829 (1992).
26. S. J. Tauster and S. C. Fung, J.Catal.,, **55**, 29 (1978).
27. D. A. Logan, E. J. Braunschweig, A. K. Dayty, and D. J. Smith, Langmuir, **4**, 827 (1988).
28. J. C. McManus and M. J. D. Low, J.Phys.Chem., **72**, 2378 (1968).

THE RAFT-LIKE STRUCTURE OF SUPPORTED PtRu₅

J.C. YANG*, S. BRADLEY**, M.N. NASHNER*, R. NUZZO*, J.M. GIBSON*
*Frederick Seitz Material Research Laboratory, University of Illinois at Urbana-Champaign, Urbana, IL 61801
**UOP, 50 East Algonquin Road, Des Plaines, IL 60017

ABSTRACT

We have examined supported PtRu₅ specimens by a variety of electron microscopy techniques, including high resolution, analytical and a novel mass-spectroscopic electron microscopy techniques. Analytical electron microscopy results showed that the relative atomic concentration of Pt to Ru for each PtRu₅ cluster is 1 to 5. The average diameter of the clusters was a 15.6Å, and the average number of atoms was measured to be 24 atoms per cluster. The combination of these techniques demonstrate that the PtRu₅ clusters are raft-like on the carbon black support.

INTRODUCTION

It is vital to understand the structure of catalytic materials on their support material, since this will clearly affect their catalytic activity. EXAFS (extended X-ray absorption fine structure) is an excellent method in determining the bonding information for the catalytic material, but it provides only averaged information. For more detailed information, transmission electron microscopy (TEM) can provide insights into the structure and chemistry as well as their distribution of supported metal clusters. We have also recently developed a mass spectroscopic technique using a scanning transmission electron microscope (STEM), which gives the number of atoms per cluster [1]. The concept of this technique is that the STEM images are due to electrons scattered at very high angles; hence, the scattering is predominantly incoherent. The robustness of this technique was verified experimentally with Re-6 clusters dispersed on graphite, where the measured number of Re atoms was 6±2 Re atoms [1]. The combination of this novel mass-spectroscopic technique with other electron microscopy techniques can provide unique insights into the 3-dimensional structure of these supported catalysts.

As an example, we are investigating a bimetallic catalyst, PtRu₅ on carbon black, by a wide variety of electron microscopy techniques. Multicomponent metal nanoclusters, such as PtRu₅, can superior catalytic behavior than single component nanoclusters. PtRu₅ on carbon demonstrates better methanol electro-oxidation in fuel cells as compared to only Pt on carbon [2]. To understand why this bimetallic catalyst is better than Pt requires knowledge of the shape of the clusters on the carbon support. In this proceedings, we present our results from a combination of these microscopy techniques, which demonstrate that the PtRu₅ clusters are raft-like on the carbon black support.

EXPERIMENT

PtRu₅ compounds were produced by a molecular precursor method [3]. To produce the TEM sample, the embedded PtRu₅ compounds in carbon black were ground with a mortar and pestle and then holey carbon grid was dipped into this powder. EDS (Energy dispersive X-ray spectroscopy) was performed on a VG (Vacuum Generators) HB601 dedicated STEM, and the HAADF (high angle annular dark field) plus secondary electron images were taken on this microscope. HREM (high resolution electron microscopy) was performed on a Hitachi 9000, equipped with a CCD camera for the images to be digitally captured. This microscope has a spatial resolution of 1.8Å.

Imaging for the STEM-based mass-spectroscopic technique was performed on a Field Emission Gun (FEG) VG HB501 STEM operated at 100 kV. The probe size used in our experiments was estimated at ~ 5 Å. Further details of the microscope operating conditions and this technique can be found in [1]. Image analysis was performed on the unprocessed digital images, where the absolute measured intensity from the clusters were converted to scattering cross-sections, which can then be converted to number of atoms. Accurate measurement of the HAADF detector efficiency is crucial for this technique. There are two methods for measuring the detector efficiency: (1). Directly measuring the current with a Faraday cup, and then tilting the electron beam onto the detector and measuring the detected signal, or (2) using the carbon as an internal standard. The data presented here was obtained by using the carbon as an internal standard. The tilted beam method was also performed in order to confirm the measurements of the number of atoms.

RESULTS

Simultaneous BF and HAADF images were taken (Figure 1(a) and 1(b)). There existed regions with virtually no clusters and regions with large number of clusters.

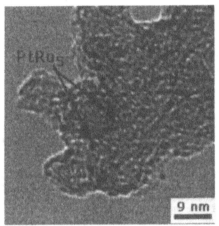 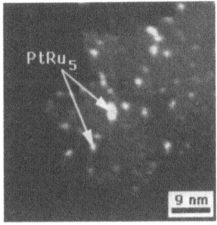

Figure 1(a): BF image of PtRu$_5$. Figure 1(b): Corresponding HAADF image

The clusters were also observed to be preferentially located at terraces, as can be seen in figure 2. Figure 2(a) is the HAADF image, which is sensitive to the atomic number, Z, of the material. Hence, the PtRu$_5$ clusters appear bright compared to the carbon support. Figure 2(b) is the corresponding secondary electron image, where the surface morphology, such as terraces, of the carbon black substrate can be seen. The arrows point to regions where the clusters are located at terraces.

EDS was performed on these specimens. 20 clusters were examined, where the relative concentrations of the Ru and Pt were: Ru mean = 77.6 ± 5% and Pt mean = 22.4 ±5%. EDS spectra taken from the entire region at 1M magnification gave: Ru mean = 77.19Å and Pt mean = 23±1.7%. Hence, we conclude that the relative atomic concentration of Pt to Ru between each PtRu$_5$ particle is constant.

Images similar to Figure 1(b) were analyzed by the STEM-based mass-spectroscopic technique. A spread in intensity is visually observable, which correlates to a spread in the measured scattered cross-sections.

The average measured scattering cross-section corresponded to 4 PtRu$_5$ groups or 24 atoms, where 30 clusters were analyzed. The diameter of these clusters were also measured from the STEM, and the average was 15.6Å. Using an averaged number density of Ru and Pt, if the shape of the cluster is spherical, then, for a diameter of 15Å, the particle contains 21 PtRu$_5$ groups, whereas for a hemispherical shape, the cluster contains 11 PtRu$_5$ groups. This data suggests that the particles are "raft-like" on the carbon black.

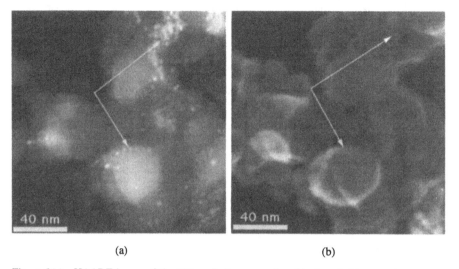

(a)   (b)

Figure 2(a): HAADF image of the PtRu$_5$ clusters on carbon black and (b) secondary electron image showing the morphology of the carbon black. The white arrows point to regions where the PtRu$_5$ clusters are at terraces.

To demonstrate this point, figure 3 is a plot of the cluster diameter vs. the number of atoms per cluster for different 3-dimensional shapes. The two theoretical fits show the number of atoms for a (a) spherical shape with diameter, d, and (b) a hemispherical shape. Clearly, since the number of atoms is considerably less than would be predicted for a hemisphere, this demonstrates that the structure of the PtRu$_5$ is oblate on the carbon black support.

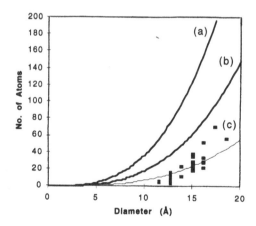

Figure 3: Plot of diameter vs. number of atoms for different shapes: (a) Sphere, (b) Hemisphere and (c) is the experimental data with the best fit where the 3-D aspect ratio is kept constant.

HREM images provide more accurate information as compared to STEM images about the diameters of the clusters. The images from a STEM is a convolution of the probe size and particle size, where the probe size is approximately 5Å, whereas the spatial resolution of the HREM is 1.8Å. However, plan-view HREM images only provide cross-sectional views of the clusters, and not information about the 3-dimensional shape. For information of the 3-dimensional shape, it was necessary to measure the number of atoms per cluster.

The PtRu$_5$ clusters was examined by HREM, where the graphite fringes of the carbon black support were used as a standard, in order to verify the diameters of these clusters. Figure 4(a) is a HREM image where the lattice fringes in the PtRu$_5$ clusters are clearly visible. To determine the crystal structure, the Fourier transform of this image was performed (figure 4(b)). The c/a ratio was measured to be 1.1±0.1 and the angle was measured to be 55±0.5°, where the values for a FCC structure is c/a = 1.155 and the angle = 54.74° for the [011] pole. This data suggests that the structure of the clusters is FCC, not HCP, which is the crystal structure for Ru. The lattice parameter was measured to be 3.6±0.3Å.

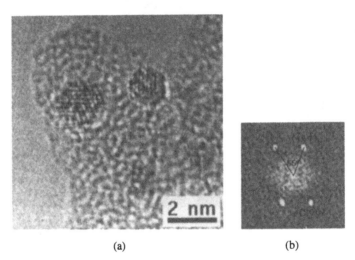

(a)   (b)

Figure 4(a). HREM image of supported PtRu$_5$ clusters. (b) Fourier transform of the HREM images, where the FFT suggests a FCC structure of the PtRu$_5$.

DISCUSSION AND CONCLUSION

Using a combination of electron microscopy techniques, including a novel STEM-based mass-spectroscopy technique, is a powerful tool for understanding the structure of real catalytic materials, i.e. PtRu$_5$ supported on carbon black. HREM provided information about the size distribution and the crystal structure of the clusters. EDS gave information about the chemistry of the clusters. 3-dimensional information was obtained by a comparison of the novel mass-spectroscopic electron microscopy technique, which gave the number of atoms per cluster, and the measured diameters. The combination of these techniques demonstrate that the PtRu$_5$ is raft-like with face-centered structure on the carbon black.

The possible reason for the raft-like structure of PtRu$_5$ is that the Pt and C do not form bonds but the Ru and C can form stable bonds. RuC exists as a compound, but there are no stable compounds with Pt and C[4]. Hence, there could be segregation of the Ru to the interface, and Pt to the surface of the clusters, as has been suggested by Nashner et al. using EXAFS [3]. Since Ru-C can form stable bonds, then this could cause the oblate structure of

$PtRu_x$. Addition of Ru could be a synthesis method to increase the amount of the Pt surface area but reduce the relative amount Pt, as compared to Pt catalysts, and thereby reducing the manufacturing costs and increasing the effectiveness of the catalysts.

There has been evidence for raft-like structures of supported metal clusters in a few other systems. For example., Prestridge et al. have reported raft-like structures of Ru using conventional transmission electron microscopy [5]. Nellist and Pennycook [6] have also reported raft-like shapes of catalytic materials, Rh on γ-Al_2O_3, by high resolution Z-contrast imaging.

The specimens used for the experiments presented in this paper were on a low-scattering support, i.e. carbon. However, this technique should also work with more realistic substrate materials, such as alumina, although the signal-to-noise ratio will worsen due to the increased scattering from the support material. To improve the signal-to-noise ratio, we are pursuing an improvement the HAADF detector efficiency. E. Kirkland and M. Thomas [7] have shown that simple mechanical modifications of the STEM can improve the detector efficiency by a factor of 100.

ACKNOWLEDGMENTS

JY thanks Dr. Ajay Singhal (Applied Materials, CA.) for many useful discussions. We thank Professor John R. Shapley (Dept. of Chemistry) for providing the organometallic compounds investigated in this research and useful discussions. This project at the University of Illinois was supported by the Department of Energy, No. DEFG02-91ER45439, and involved extensive use of the facilities within the Center for Microanalysis of Materials of the Frederick Seitz Materials Research Laboratory. We are particularly grateful to Peggy Mochel and Ray Twesten for assistance with the STEM. Also, AT&T/Lucent Corporations, especially Steven Berger and William Brinkman, are kindly acknowledged for their donation of the VG-STEM instrument to the University of Illinois.

REFERENCES

1. A. Singhal, J.C. Yang and J.M. Gibson, Ultramicroscopy. **67**, 191-206 (1997).

2. V. Radmilovic, H.A. Gasteiger and J. P. N. Ross, Journal of Catalysis, **154**, 98-106 (1995).

3. M. S. Nashner, A. I. Frenkel, D. L. Adler, J. Shapley and R. Nuzzo, J. of Am. Cer. Soc., **to be published** (1997).

4. M. Hansen, Constitution of Binary Alloys. 2 ed. McGraw-Hill (1958).

5. E. B. Prestridge, G.H. Via and J.H. Sinfelt, J. of Catalysis. **50**, 115-123 (1977).

6. P. D. Nellist and S.J. Pennycook, Science. **274**, 413-415 (1996).

7. E. J. Kirkland and M.G. Thomas, Ultramicroscopy. **62**, 79-88 (1996).

ON EXTENDED X-RAYS ABSORPTION FINE STRUCTURE CHARACTERISATION OF IRON PLATINUM CLUSTERS

G. D'Agostino[1], A. Filipponi[2], A. Verrazzani[3] and G. Vitulli[3]

[1] Divisione Nuovi Materiali, ENEA, Centro Ricerche Casaccia - Roma (Italy)

[2] Beam Line 18, ESRF, Grenoble (France)

[3] Dip. di Chimica e Chimica industriale, Università degli studi di Pisa and Centro di Studi del CNR per le Macromolecole stereordinate ed otticamente attive, V. Risorgimento 35 PISA (Italy)

ABSTRACT

Present paper reports on Extended X-rays Absorption Fine Structure (EXAFS) Spectroscopy characterisation of Fe/Pt clusters. Samples of different nominal iron contents (FePt4 and FePt2) have been prepared and characterised by direct measurements of absorption coefficient at Pt L_3 and Fe K edges. Catalysts were prepared by precipitation from suitable mixing (in the appropriate stoichiometry) of mesitylene suspentions containing small clusters of either platinum or iron. Single species suspention, in turn, were prepared by vapor metal deposition.

Catalitic activity of prepared samples was tested on cinnammic aldehyde. As a general result, platinum clusters exhibit activity in the hydrogenation of both carbon double bond and aldehyde group with a preference for the latter. Iron addition to the pure platinum catalyst reverts the selectivity while leaving the catalitic activity almost unchanged.

The present paper represents a temptative to achieve structural information on Fe and Pt short range environments, thus providing insights on the dynamics of the catalytic process.

INTRODUCTION

Interest in nanoscale system has been growing fast during last decade. Both single particle and aggregates are appealing for structural and functional applications [1]. Nanometric metallic clusters exhibit a large surface to volume ratio, thus enhancing their species natural catalytic activity [2]. In particular, platinum represents a natural catalyst for hydrogenation process. Platinum clusters are also appealing for optical applications as their addiction to inert matrices may enhance the resulting non linear optical response.

Among the established methods of cluster preparation, the technique which is variously known as metal atoms or metal vapor synthesis is taking a relevant position [3-4]. The former technique has been employed to prepare stable suspensions of solvated Fe and Pt small clusters. Starting from former precursor clusters, composite Fe/Pt catalists have been prepared.

Lattice structure of powder and its macroscopic morphology have been studied by means of TEM (Transmission Electron Microscope) and HRTEM (High resolution TEM) measurements [5]. Since electron microscopy is a non chemical selective tool, it was not easy to achieve information on Fe laying in global structure of the sample. In order to obtain different structural information on Fe and Pt species, EXAFS measurements have been performed at the two different Pt L3 and Fe K edges. By this means one may obtain separate information on Fe and Pt structural short range environment and inferring possible dynamics for the catalized hydrogenation process. As it will be clarified later, presence of non trivial tungsten contamination complicates the EXAFS characterization.

SAMPLE PREPARATION

Samples have been prepared in two steps. First process is a typical vapor deposition: each species (Fe and Pt) has been evaporated in an arene (toluene or mesitylene) saturated atmosphere and deposited (together with arene) on reactor cold surface (kept at -196°C). Solid deposit was then warmed up to some -40°C, thus leading to a stable suspension of extremely small clusters. The resulting precursor suspentions have then been mixed, in the appropriate stoichiometry (FePt4 or FePt2), and wormed up to room temperature. Upon heating, the clustering process initiates and the resulting larger particles are slowly deposited. Extracting and drying of deposit provides the final polycrystalline powder.

By means of the technique above, samples of two different compositions were prepared. In the forthcoming they will referred to as FePt4 and FePt2, lending names from their stoichiometric mixing. Non chemical notation is selected to stress the variation with respect to the formation of actual $FePt_4$ or $FePt_2$ compounds. Silica supported samples were also attained by addiction of pure silica to the solution before the worming process.

Scanning Electron Microscopy (SEM) characterization provided evidence for a sponge-like structure with typical grain size of some hundred nanometer. Similarly, High Resolution Transmission Electron Microscopy (HRTEM) allowed to estimate the particulate dimensions as some 3-4 nm [5].

CATALYTIC ACTIVITY

Pt clusters prepared by the former method exhibit a noticeable catalytic activity in the hydrogenation process. In the present paper it was tested by means of the hydrogenation of the cinnamic aldeheyde. As a general result, Pt catalysts activate the hydrogenation of both carbon double bond and carbon oxygen double bond with a preference for the former reaction. In the present case, basically, two competitive processes are allowed (showed in fig. 1).

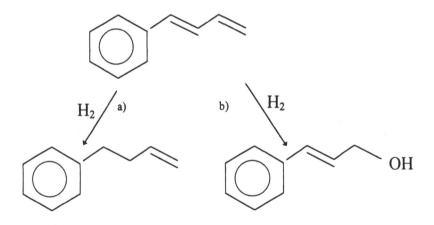

Fig. 1 *Hydrogenation of cinnaldehyde in H_2 reducing environment. Main two reactions are labelled as a) and b).*

Selectivity of pure Pt prepared samples exhibited a branching ratio for the channels a) and b) of about 2; whereas Fe doped samples inverted their selectivity in favor of the double bond hydrogenation with a branching ratio of .25 (i.e. b vs a channel frequency is about 4).
Selectivity did not exhibit the expected dependence on nominal composition.

ABSORPTION MEASUREMENTS

One of the most appealing features of EXAFS spectroscopy is represented by its chemical selectivity. By measuring the absorption coefficient at the energy following the different absorption edges one may probe the short range structural environment of each species. Presently, the most interesting regions are around Fe K Edge at 7.1108 KeV and Pt L3 Edge at 15.456 KeV.

Fig.2 shows the absorption coefficient of FePt4 sample measured in a wide range of frequencies. As indicated by the arrow, Pt L3 absorption edge exhibits a small pre-edge due to W L2 absorption edge. Presence of W is also evidently confirmed by its L3 absorption edge. Tungsten represents a relevant contamination which can be estimated (by the height of its L3 signal) at few wt %. Source for W contamination is the solenoid crucible from which Pt is evaporated. Unless quantitatively non negligible, W contamination does not seem to interfere with the hydrogenation activity or prepared samples. Analysis of L3 W absorption edge is consistent with the hypothesis that most of W is bond to C. One expects such W-C bond to be mainly formed during the evaporation process.

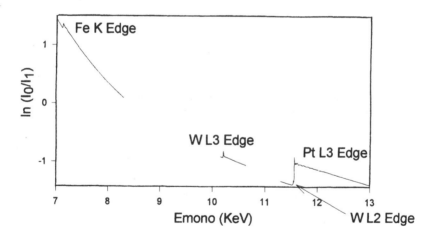

Fig. 2 Absorption coefficient profile as a function of the monochromator energy. Pt L3 and Fe K edges are scanned together with L2 Edge of tungsten that resulted to be a relevant contaminant. Fe K and Pt L3 jumps are consistent with nominal composition.

Absorption by W at its L2 edge slightly influences Pt L3 signal. To prevent against possible misleading interpretations of Pt L3 EXAFS profile, W L2 Absorption signal was subtracted by assuming a proportionality with W L3 correspondent structural signal. This is quite correct as the two edges differ only for the initial total angular momentuum. Result of such subtraction is showed in fig. 3. As can be seen, a residual pre-edge deviation survives the subtraction. Such residual component of L3 W signal is probably related to slight differences in the density of states near the fermi level and does not influence the EXAFS region of Pt L3 absorption Edge.

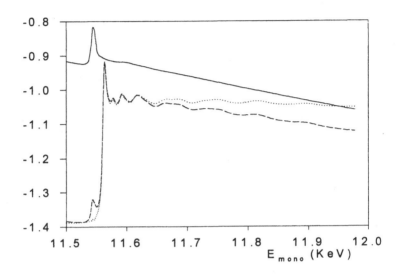

Fig. 3 *a Subtraction of W L2 edge from Pt L3 edge. Some residual pre-edge difference is left due to slight differences in density of states before fermi level.*

Fig.4 shows the $\chi(k)$ signal of FePt4 sample at Pt L3 absorption edge compared with that of a Pt reference foil. As can be seen, prepared sample exhibits only slight deviations form ordinary polycrystalline material. Such deviations are presently ascribed to presence of substitutional iron (or possibly W) in fcc Pt matrix. Deviations related to the small dimensions of clusters are expected to occur at lower grain-size (some nm) [7]. Standard analysis (by means of GNXAS package [8]) limited to the first shell signal indicate r=2.78(1) for the first neighbors distance that is consistent with a pure Pt fcc system. Since relevant addition of iron in Pt fcc structure should have changed the lattice parameter of some per cent, one expects the effective Fe contents in Pt clusters to be realistically less than the 1:4 ratio of the nominal composition. In fact, as will be clarified later, majority of Fe is bonded to carbon.

The fine structure signal ($\chi(k)$) of Pt L3 edge as a function of the photo-electron momentuum is compared with a Pt reference foil one in fig.4. The differences in the signal are mainly ascribed to presence of some substitutional Fe (or W) in the Pt matrix.

Inspection of Fe K absorption edge, provides some qualitative information on the Fe short range environment. First the appearance of a pre-edge signal at 7.111 KeV and the (related) shift of Fe K edge at 7.126 represents a clear indication that the density of states (dos) before fermi level is very different from that of a bcc (body centered cubic) Fe. Moreover the shape of the $\chi(k)$ signal (fig.5) is completely different from that of a bcc Fe (or from that of iron carbides such

118

as Fe₃C) and drastically damped. Furthermore a first shell best fit analysis provides r=1.85Å as the most likely Fe-C first neighbors peak. All the previous observations point toward the formation of some metal-organic compound [6].

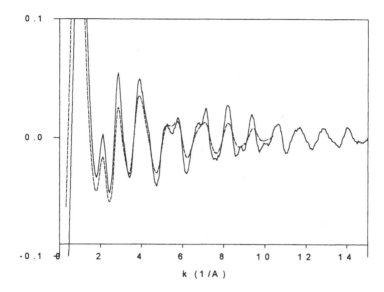

Fig. 4 *Dashed line represents EXAFS signal at Pt L3 absorption edge for FePt4 sample. Integer line represents Pt foil reference signal. Differences are ascribed to Fe (or W) substitutional defects. Photo-electron momentum is expressed in (1/Å)*

In principle one could try to reconstruct the Fe K edge EXAFS signal as a superposition of two signals coming from both Fe-Pt bond in the Pt matrix and Fe-C of some test compound. High level of noise, due to the presence of Pt absorption ground and the intrisic complexity of heavy transition metals spectra, does not allow to perform such a program. However one may exclude conspicuous formation of typical Fe-Fe or Fe-Pt bonds.

EXAFS signals at both Fe and Pt absorption edges of FePt2 sample coincide with those of FePt4 sample within measure uncertainty. This is consistent with catalytic activity measurements which seem to be also insensitive to nominal stoichiometry. A possible interpretation is that only a very small amount of Fe is trapped in the Pt fcc matrix during the cluster growth, thus leading to close similar EXAFS spectra.

CONCLUSIONS

EXAFS measurements allowed characterization of iron doped Pt catalysts prepared from vapor deposited precursor clusters. Evidence for formation of the typical fcc short range environment for platinum atoms was provided in consistence with HRTEM measurements. On the other hand, measurements at iron K edge indicate formation of typical Fe-C bonds and allow to exclude conspicuous presence of Fe-Fe of Fe-Pt bonds. Therefore iron and platinum do not seem to belong to the same lattice.

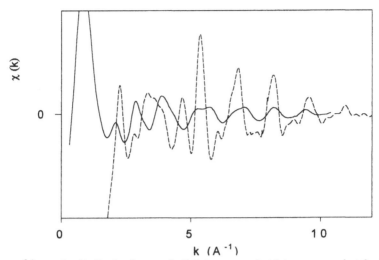

Fig. 5 $\chi(k)$ signal at Fe K edge for sample FePt3 compared with it correspondent for Fe reference foil. The observed profile suggest formation of Fe-C organic compounds.

As a result of the former considerations, the most realistic interpretation for the inversion of catalist selectivity upon addition of iron to platinum clusters seems the following. When cinammic aldehyde approaches the catalyst it tends to direct its oxygen toward iron thus reducing the probability for the aldehydic group to approach platinum and, consequently, inhibiting the catalytic action for the relative reaction.

ACKNOWLEDGEMENTS

Work was partly supported by European Synchrothron Radiation Facility (ESRF).

REFERENCES

[1] C. Hayashi, Physics Today 40, 44 (1987).
[2] G. Schmid, Chem. Rev., 92 1709 (1992).
[3] M.P.Andrews and G.A. Ozin, Chem. Mater. 1,174 (1989).
[4] K.J. Klabunde,Y.Y.X. Lia and B.J. Tan, Chem. Mater. 3, 30 (1991).
[5] G.Vitulli, E. Pitzalis, R. Lazzaroni, P. Pertici, P. Salvadori, O. Salvetti, S. Coluccia and G. Marra, Mat. Science Forum 15, 93 (1995).
[6] S. Della Longa, G. Amiconi, I. Ascone, A. Bertollini, A. Bianconi and A. Congiu Castellano, J. Phys. IV France 7 C2-629 (1997).
[7] G. D'Agostino, A. Pinto and S. Mobilio Phys. Rev B 48, 14447 (1993).
[8] A. Filipponi and A. Di Cicco, Phys. Rev. B 52, 15135 (1995).

ADVANCED METAL/CERAMIC CATALYSTS FOR HYDROGEN GENERATION BY STEAM REFORMING OF HYDROCARBONS

S.F. TIKHOV, V.A.SADYKOV, A.N.SALANOV, Yu.A.POTAPOV, G.S.LITVAK, S.V.TSYBULYA, and S.N.PAVLOVA

Boreskov Institute of catalysis of the SB RAN, Pr. Lavrentyeva, 5, Novosibirsk, 630090, RUSSIA.

ABSTRACT

The main features of porous metal/ceramic catalysts for hydrocarbons steam reforming their formation via hydrothermal treatment of the powdered aluminum and lanthanum nickelate mixture in the confined space are elucidated. For catalysts obtained via this route, their polymodal pore structure was found to approach the optimum level. The phase composition of the active component precursor was shown to have a great impact on the specific catalytic activity of catalysts.

INTRODUCTION

Methane steam reforming is currently one of the main industrial processes for hydrogen generation [1]. Earlier [2], monolith catalysts based upon powdered lanthanum nickelates and shaped using aluminum hydroxide as a binder were shown to be more active and stable in the reaction of methane dry reforming than traditional catalysts prepared via the precipitation route. Recently, we have developed a new method for a porous Al/Al_2O_3 cemet synthesis, which can be used as a support or encapsulating matrix to fix the active components [3]. In this work, we present results on synthesis of the methane steam reforming catalysts based upon perovskites encapsulated within such cemet matrix along with studies of the main features of their pore structure formation as related to activity.

EXPERIMENTAL

As starting compounds, powdered Al of the PA-4 grade, $LaNiO_3$ synthesized either by the high-frequency induction plasma thermolysis of the mixed nitrates solution [4] («plasmochemical sample»), or by thermal decomposition of the same solution under air in a crucible with a subsequent calcination at 1100 °C («ceramic sample») were used. Cemet synthesis procedure includes mixing of powders and filling by this mixture into a specially designed die (press-form) constructed in such a way as to ensure a free access of a water vapor through some channels while firmly holding the compacted powders. After the autoclave treatment at 200 °C for 3 h under the saturated water pressure, a monolithic cemet thus formed was removed from the die, dried and calcined at 800 °C for 2 h [3]. The active component content in the cemet was in the range of 30-40%.

Catalytic activity was determined by the degree of CH_4 conversion measured at 750 °C using a batch-flow catalytic installation (recirculation rate ~ 1000 l/h) with the reaction mixture $CH_4/H_2O = 1:2$ feed at a velocity of 2 ml/s at atmospheric pressure into the reactor equipped with 0.5 g of catalyst diluted by quartz in a 1:10 ratio. Before measurements, the catalysts were activated in a stream of hydrogen with a temperature ramp from ambient temperature to 750 °C within 0.5 h and keeping this temperature for 1 h.

The particle size distribution was investigated by the Coulter method using a TA-2 machine and DAS computer [5]. Specific surface area was determined by the BET method using Ar thermal desorption data. The pore size distribution was estimated from the mercury porosimetry data using an AutoPore 9200 machine. The thermal analysis of samples predried at 120 °C was performed in a steam of air using a Q-1500 derivatograph. The phase composition was analyzed with an URD-63 diffractometer using CuK$_\alpha$ radiation. The morphology and the textural features of samples were studied by scanning electron microscopy using a BS-350 «Tesla» machine.

RESULTS AND DISCUSSION

Precursor properties

Fig.1 shows a particle size distribution of the powdered components used in the synthesis. As follows from these data, for powdered Al, particles with diameters in the range of 20-25 μm predominate. For «plasmochemical» perovskite sample, a mean size of particles aggregates is smaller being in the range of 6-8 (m. A broad particle size distribution approaching a bimodal one was observed for the «ceramic» perovskite.

According to XRD data, all three types of the most strong reflexes typical for the fcc lattice, were observed for powdered Al. A phase of the hexagonally distorted LaNiO$_3$ perovskite with small admixtures of the phase with a layered-type structure (La$_2$NiO$_4$) and NiO oxide were revealed in the ceramic sample. For «plasmochemical» sample, the layered-type perovskite phases were the most abundant along with the oxides of lanthanum and nickel.

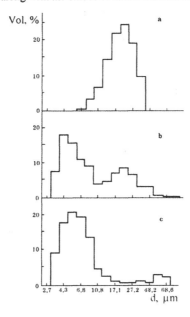

Fig. 1. Particle size distribution in different powders: a-Al, b-La-Ni-O ceram, c-La-Ni-O plasm.

Composite catalyst formation

During hydrothermal treatment, a partial oxidation of Al and formation of boehmite particles on its surface occurs [3]. The resulting increase of the solid phase volume in a confined space causes a self-pressing of powders into a monolith. Fig. 2 demonstrates the thermal analysis data for catalysts samples with the $LaNiO_3$-based active component. For comparison, the data for a pure cermet matrix are also shown here. For all samples, endoeffects at 120 ° C and 520 °C accompanied by a weight loss were observed and assigned to desorption of water and boehmite decomposition, respectively [3]. In all cases, exoeffect at ~ 330 °C was also observed. Earlier, [3], it was assigned to a structural relaxation in the metastable aluminum oxohydroxide phase formed by the hydrothermal oxidation of aluminum. More pronounced exoeffects for samples containing lanthanum nickelates imply that nickel and lanthanum cations are in part incorporated into the aluminum oxohydroxide lattice thereby increasing is disordering. A second exoeffect situated at ~ 640 °C is assigned to aluminum oxidation. It is superimposed on a large endoeffect at 660 °C caused by the melting of aluminum. At higher temperatures, a continuous weight increase due to the melt oxidation is observed. The last phenomenon was much less pronounced for the perovskite - containing composites, indicating a retardation of the aluminum oxidation due to the presence of Ni and La in the oxidic layer encapsulating the melt droplets. For perovskite - containing systems, two new endoeffects accompanied by a weight loss were revealed at ~ 350 °C and ~385 - 400 °C. These features can be assigned to decomposition of lanthanum hydroxide and/or mixed nickel-lanthanum oxohydroxides, which usually decompose at temperatures intermediate between those corresponding to individual hydroxides [7,8].

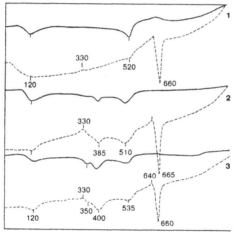

Fig. 2. Differential thermal analysis data of powdered precursor after HTT treatment and drying at 120 °C: 1 - Al, 2 - Al⁺(La-Ni-O plasm), 3 - Al⁺(La-Ni-O ceram). Full curve - DTG, dashed curve - DTA.

After calcination at 800 °C, the intensity of the X-ray reflexes corresponding to $LaNiO_3$ phase increases while that of the NiO and La_2NiO_4 phases falls, remaining though rather high for catalysts with the «plasmochemical» perovskite. In all samples, phases of metallic Al and γ-alumina were also present. Stabilization of the low-temperature modification of alumina observed here is clearly due to a well-known effect of lanthanum cations, which is also manifested in the high surface area of the catalysts (see Table 1).

3.   Catalyst pore structure
    For catalysts calined at 800 °C, the texture appears to be of a mixed corpuscular-sponge character, where regions with a crust formed by solidification of a molten aluminum are followed by a regions of densely packed sintered particles. For composites with the «plasmochemical» active component, the nearly ideal hollow spheres of the parent powder were rather well preserved being sintered into a sponge-like structure (Fig. 3a). Such particles are not typical for samples with the «ceramic» perovskite (Fig. 3b).
    Fig. 4 shows a typical dependence of the cumulative pore volume versus their radii for two types of catalysts studied here. In both cases, pore size distribution is of a polymodal type. For composite with the «plasmochemical» active component, a component, a cumulative pore volume is higher due to a bigger volume of ultramacropores ($\sim$ 10-50 μm) as well as mesopores (0.01-0.1 μm). As judged from the specific surface value (Table 1), composites also have numerous micropores persisting even after high-temperature calcination, which could not be revealed by the mercury porosimetry.
    To analyze the pore structure formation, we should take into account that a mean diameter of the Al particles changes only slightly after oxidation. Hence, to a first approximation, all specific features appear to be mainly determined by a mode of powder particle packing while filling the press-form (die) [9]. Thus, a smaller pore volume for composite with the «ceramic» perovskite can be assigned to an optimum ratio between fractions of the large and small particles, despite a great fraction of the former entities. Hence, a kind of a «dense packing» seems to be favored with small particles filling the voids between the more bulky aggregates [11]. An excess of a fine fraction for «plasmochemical» perovskite is clearly not favorable for such a dense packing. Further, in the latter case, macroporosity can be enhanced due to internal volume of hollow spheres constituting a great part of the plasmochemical sample particles.

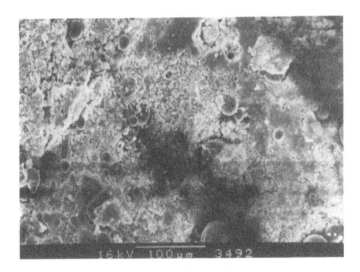

Fig. 3a. SEM electron micrographs of obtained composites after heat treatment at 800 °C with La-
    Ni-O plasm.

Fig. 3b. SEM electron micrographs of obtained composites after heat treatment at 800 °C with La-Ni-O ceram.

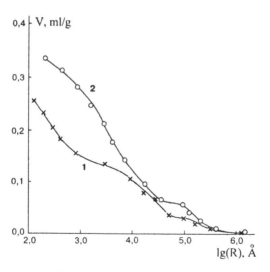

Fig. 4. Cumulative pore volume of composite catalysts vs. pore radii: 1 - with La-Ni-O ceram, 2 - with La-Ni-O plasm.

A second factor that might affect composites pore structure is aluminum transfer between various particles of these heterogeneous multiphase systems. During hydrothermal oxidation, such transfer can proceed via dissolution in the water condensed in pores with a subsequent precipitation at some other places including surface of perovskite particles. In this case, the driving force for such a transfer is the chemical potential difference due to a different chemical composition and a particles surface curvature. In the course of high temperature treatment, the mass transfer can be generated due to a melting of aluminum and its oxidation. In both cases, the mass transfer is expected to fill the space between contacting particles thus being the most pronounced for composites containing «plasmochemical» perovskite with a large fraction of smaller particles and a higher surface curvature. For this sample, filling the space between the perovskite particles should decrease the share of pores ~ 1 μm and increase the fraction of pores with radii ~ 0.01 - 0.1 μm, that is indeed observed. In all catalysts, microporosity can be assigned to slit - shaped pores within the alumina particles formed in the course of boehmite dehydroxylation [3].

4.   Catalytic properties

As follows from Table 1, activity and CO selectivity are much higher for catalysts containing «ceramic» perovskite as a precursor. For a catalyst with the «plasmochemical» active component, its loading is comparable being only slightly (20 - 25 rel. %) less a factor that could not account for a much lower activity. In this case, an even higher porosity and a great fraction of the transport pores are not capable of ensuring good activity. Hence, the most important factor affecting activity of these systems appears to be the phase composition. Thus, for the ceramic sample, the LaNiO$_3$ phase predominates, and after hydrogen activation it is converted into the well-dispersed metallic Ni particles stabilized against sintering in the lanthana matrix. For the plasmochemical sample, a part of the Ni component remains non-reduced being confined within the layered La$_2$NiO$_4$ particles, a part forms bulky Ni particles via reduction of the NiO particles and are susceptible to subsequent sintering. Another possible explanation - surface blocking by carbonacios species, absorbed on lanthanum ions, which leads to stabilization of cationic forms of nickel.

However, pore structure also seems to play an important role for methane steam reforming on these catalysts [12]. Thus, for the GIAP-16 catalysts (26 wt.% NiO/(-Al$_2$O$_3$), methane conversion extrapolated to our conditions is ~ 83%, that is rather close to the value for the ceramic sample (Table 1). But, for GIAP-16 catalyst conversion drops sharply for granules, while it is less sensitive to the granule size of our catalyst.

Table 1. Various properties of the catalysts.

No	Active Component wt%	Catalyst fraction mm	Specific Surfacearea m^2/g	Catalytic Properties		
				X_{CH_4} (%)	K^*, $\frac{ml\ CH_4}{g\ s\ atm}$	S_{CO} (%)
1	La-Ni ceram., 40	0.5 - 1.0	34	81.1	17	82.1
2	La-Ni ceram., 40	3 - 5	34	66.2	8	76.8
3	Ni(GIAP-16), 26	0.25 - 0.5	-	~ 83	20	-
4	Ni(GIAP-16), 26	3 - 5	-	~33	0.2	-
5	La-Ni plasm., 40	0.5 - 1.0	40	4.7	2	36.3

* First order efficient rate constant

CONCLUSIONS

Porous composite materials based upon Al/Al_2O_3 cermets are shown to be promising as methane steam reforming catalysts. Peculiarities of their pore structure formation were studied. Phase composition of the active component perovskite - type lanthanum nickelate precursor affecting metallic nickel dispersion and stability was shown to be the most important parameters for high activity.

ACKNOWLEDGMENTS

We thank V.B. Fenelonov for fruitful discussions.

REFERENCES

1. J.P.Rostrup-Nielsen in Catalysis: Sci.@Tech., edited by J.R. Anderson and M. Boudart (Akademie-Verlag-Berlin 5, 1984), p. 1-118.
2. L.A. Isupova, V.A. Sadykov, S.F. Tikhov, O.N. Kimkhai, O.N.Kovalenko, G.N.Kustova, I.A. Ovsyannikova, Z.A. Dovbii, G.N. Kryukova, A.Ya. Rozovskii, V.F. Tretyakov, V.V. Lunin, Catal. Today, **27**, p. 249 (1996).
3. S.F. Tikhov, A.N.Salanov, Yu.A.Palesskaya, V.A.Sadykov, G.N.Kustova, G.S.Litvak, N.A.Rudina, V.A.Zaikovskii, S.V.Tsybulya, React.Kinet.Catal.Lett. (1998) [accepted].
4. V.D.Parkhomenko, P.N.Tsybulev, Yu.I.Krasnokutskii, Plasmochemical Technology, Vischa Shkola, Kiev, 1991, pp. 99-102 (in Russian).
5. T.Alem, "Particle Size Measurement", Capt.Hill, London, 1981, p. 590.
6. Fenelonov V.B. React.Kinet.Catal.Lett., **52**, 367 (1994).
7. P.V. Klevtsov, L.P.Sheina, Izv.AN SSSR.Neorg.Materiali, **1**, p. 912 (1965) [in Russian].
8. V.P.Chalii, Metal hydroxydes, Naukova Dumka, Kiev, 1972, p.94 [in Russian].
9. R.Ya.Popilskii, F.D.Kondratov, Pressing of the ceramic materials, Metallurgiya, Moscow, 1968, p.63 [in Russian].
10. R.K.McGeary, J.Amer.Ceram.Soc., **44**, p.513 (1961).
11. V.B.Fenelonov, D.V.Tarasova, V.Yu.Gavrilov, Kinetika I kataliz, **19**, p.222 (1978) [in Russian].
12. N.N.Bobrov, I.I.Bobrova, V.A.Sobyanin, Kinetika I kataliz, **34**, p. 686 (1993) [in Russian].

FORMATION OF PLATINUM AND PALLADIUM BIMETALLIC SUPERFINE PARTICLES BY ULTRASONIC IRRADIATION

Toshiyuki Fujimoto*, Shinya Terauchi*, Hiroyuki Umehara*, Isao KOJIMA*
and William Henderson**
*National Institute of Materials and Chemical Research, Tsukuba, Japan
** on leave from the University of Waikato, Hamilton, New Zealand

ABSTRACT

Pt/Pd bimetallic nanoparticles were prepared by ultrasonic irradiation of aqueous solutions containing salts of both metal ions. Uniform bimetallic particles were found to have a 2.8 nm average diameter and a narrow size distribution. The size was similar to Pt monometallic particles, and the shape was similar to Pd monometallic particles, both prepared by ultrasound in the same way.

INTRODUCTION

Novel metal nanoparticles have been used in many fields such as catalysis and optics. They also have potential application in quantum electronics devices. In order to design high performance materials, it is of paramount importance to control particle sizes and structures. Metal nanoparticles which are derived from two metal species show different catalytic activities from the mixture of single-metal nanoparticles. Some preparation methods [1,2,3] and promotions of activities and/or selectivities using bimetallic nanoparticles have been reported [1,4].

On the other hand, ultrasound has become an important tool in chemistry. When strong ultrasound is irradiated to solutions, bubbles in solution are implosively collapsed by acoustic fields, and high temperature and high pressure fields are produced at the center of bubbles. This effect is known as an acoustic cavitation. The temperature of the hot spot has been estimated [5] to be over 5000K which is high enough to decompose molecules in bubbles. Suslick et al. have applied this phenomenon to prepare amorphous iron from $Fe(CO)_5$ [6]. Formation of Ag [7] and Au [8] nanoparticles using ultrasonic irradiation of aqueous solutions have been reported. Very recently, formation of Au/Pd bimetallic particles was also reported [9]. Here we report formation of Pt and Pd bimetallic nanoparticles using ultrasound.

EXPERIMENTAL

$H_2PtCl_6 \cdot 6H_2O$ and K_2PdCl_4 were dissolved in argon-saturated water to make $ca.$ 3 mmol·dm^{-3} solutions. The solutions were mixed together with argon-saturated water to make five sample solutions which consisted of different Pt:Pd molar ratios (1:0, 4:1, 1:1, 1:4 and 0:1, which were denoted as S(1:0), S(4:1), S(1:1), S(1:4) and S(0:1), respectively). For all solutions the total metal ion concentration was controlled to 1 mmol·dm^{-3} and poly(N-vinyl-2-pyrrolidone) (PVP, K-30,

MW 40000) added as a protecting group. The concentration of PVP was *ca.* 0.1 mmol•dm^{-3}.

The metal salt solution (24 ml) and 1 ml of argon-saturated ethanol were put into the cylindrical sonication reactor (glass made, 25 mm diameter) which was placed in a water bath. Because the local heating produced by the cavitation strongly depends on the solution temperature [10], the temperature of the water bath was held at 278 ± 0.5 K during sonications. Ultrasonic irradiations were made with a collimated 20 kHz beam from a ceramic transducer with titanium amplifying horn (Sonics and Materials, VCX-600) directly immersed in the solution. The power of the ultrasound was 100 W.

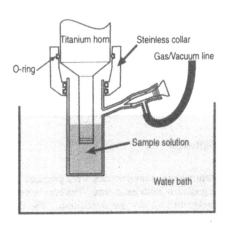

Figure 1 Schematics of the Sonication Reactor

The sonication reactor and titanium horn were connected by an O-ring. A schematic of the reactor is represented in Figure 1. Since the reactor has a teflon valve, the solutions can be treated without exposing to air by standard Schlenk techniques. The irradiation was stopped every minute and UV-VIS spectra were observed using a UV-VIS2000 (Shimazu) instrument. Some drops of the solutions were placed onto a carbon support film stuck on a copper microgrid, and vacuum dried. Transmission electron microscopy was performed using a JEOL2000FXII. The sizes of 500 particles were measured on micrographs in order to obtain averaged diameters and size distributions. All manipulations were performed under an argon atmosphere except when introducing the microgrids to the electron microscope.

RESULTS

Metal ions can be reduced by radicals produced by acoustic cavitation. The expected sequence of reactions that would lead to the formation of novel metal nanoparticles are shown in equations (1) to (6) [7,11]

$$H_2O \qquad \text{sonication)))))} \qquad H^\bullet + HO^\bullet \qquad (1)$$

$$ROH \qquad \text{sonication)))))} \qquad H^\bullet + HO^\bullet + R^\bullet + RO^\bullet + O \qquad (2)$$

$$RH + HO^\bullet \qquad \rightarrow \qquad R^\bullet + H_2O \qquad (3)$$

where RH *denotes PVP or ethanol*

$$nH^\bullet + M^{n+} \qquad \rightarrow \qquad M + nH^+ \qquad (4)$$

$$nR^\bullet + M^{n+} \qquad \rightarrow \qquad M + R' + nH^+ \qquad (5)$$

where R' + H = R

$$n(M) \qquad \rightarrow \qquad (M)_n \qquad (6)$$

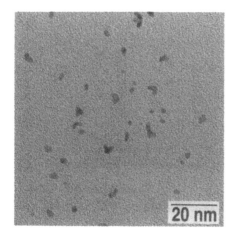
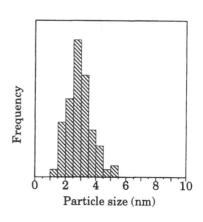

Figure 2 Electron micrograph and particle size distribution of
Pt nanoparticles prepared from S(1:0)

The clear yellow-colored solution of S(1:0) changed to very pale grayish-yellow and then turned to dark brown upon ultrasonic irradiation. In the UV-VIS spectrum, absorption bands around 375 nm and below 300 nm firstly decayed (0-50 minutes) and after that an increase of the base line was observed during the sonication. Only a few Pt particles larger than 2 nm were observed in electron micrographs of the sample prepared from the solution after the first stage of sonication. This result indicates that the consumption of $PtCl_6^{2-}$ predominantly occurred at an early stage of the sonication, and the reduced Pt atoms and/or small clusters grow to colloids. The solution color became darker by further sonication, and the point at which no further change of UV-VIS spectra was observed was chosen as the reaction finish. Figure 2 shows an electron micrograph and particle size distribution of Pt nanoparticles after sonication. Irregular particle shapes are found, including spherical, football-like, distorted three pointed star shapes, and so on. The average diameter is 2.9 nm and the standard deviation of the particle size is 0.81 nm.

In the case of S(0:1), the original solution (clear yellow-orange) showed a similar reaction sequence to S(1:0), but the first-stage reactions were completed in 5 minutes. These results suggest that the formation of the Pd particles was much faster than that of Pt under the same conditions. The absorption bands at 310 nm and 425 nm which were characteristic of S(0:1) before sonication almost disappeared and the increase of the UV-VIS spectrum base line was observed after 5 minutes sonication; reaction was complete after 10 minutes sonication. Figure 3 shows an electron micrograph and particle size distribution of Pd nanoparticles after sonication. The Pd particles

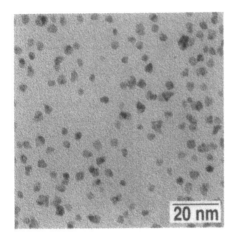
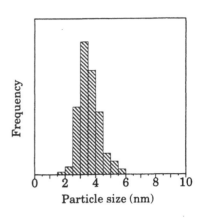

Figure 3 Electron micrograph and particle size distribution of
Pd nanoparticles prepared from S(0:1)

were observed as larger and regular shaped particles compared with the Pt particles. The average diameter is 3.6 nm and the standard deviation of the particle size is 0.69 nm. Although edges are observed for some particles, most are spherical.

For the mixture of Pd(II) and Pt(IV) it was expected that the formation of the Pd particle would occurred initially and then a Pt outer shell might be formed, which is similar to the seed-growing method[4]. However, the results of sonication of the mixture were notably different. Figure 4 shows the spectral change of S(1:1) during sonication. Absorptions around 375 nm and below 300 nm [which were characteristic of S(1:0) before sonication] quickly decayed upon ultrasonic irradiation. The spectral change is very similar to that of S(1:0) at the early stage but the reaction rate was similar to that of S(0:1). After 5 minutes irradiation, the characteristic absorption of S(1:0) almost disappeared and the spectrum shape was similar to the spectrum of S(0:1)

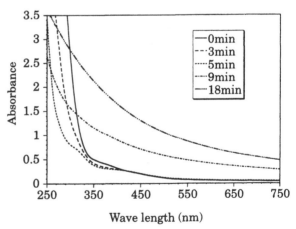

Figure4 Spectral change of Pt/Pd aqueous solution
S(1:1) during sonication.

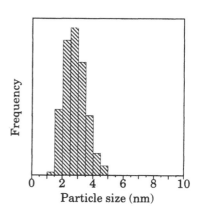

Figure5   Electron micrograph and particle size distribution of
Pt/Pd bimetallic nanoparticles prepared from S(1:1)

before sonication.   These results indicate that reduction of Pt(IV) predominantly occurs in the
mixtures initially.   Nagata et al. reported a similar phenomenon for the Au/Pd system [9] and
suggested that Au-containing ions were reduced by intermediate valence ions of Pd or Pd atoms
due to the difference of their redox potentials.   Although the difference of the redox potentials in
this system ($PtCl_6^{2-}$ / $PtCl_4^{2-}$: 0.68V, $PtCl_4^{2-}$ /Pt:0.73V, and $PdCl_4^{2-}$ /Pd: 0.63V) is smaller than
that of the Au/Pd system, reduction of Pt ions by intermediate valence ions of Pd or Pd can occur
during sonication.

The color of the solution turned to dark brown by further sonication.   The spectra of the colloid
solutions from the mixture were similar to Pt/Pd bimetallic colloidal solutions previously prepared
by thermal reduction of the metal salt solutions, and were different from the spectrum of a mixture
of Pt and Pd colloidal solutions prepared by ultrasound.   These results suggest that Pt/Pd bimetallic nanoparticles were formed by the ultrasonic irradiations of the S(1:1) mixture.   Electron micrographs and particle size distributions of the Pt/Pd bimetallic nanoparticles prepared from S(1:1) are
shown in Figure 5.   The average diameter and standard deviation of the Pt/Pd bimetallic nanoparticles
are 2.8 nm and 0.68 nm, respectively.   Most particles are observed to have a spherical shape.   It is
notable that the particle shape is similar to the Pd nanoparticles whereas the particle size is similar
to the Pt nanoparticles.   This result is also considered to be evidence for formation of Pt/Pd bimetallic particles by ultrasonic irradiation.   From the results of the spectral change and electron
microscopy, it is suggested that (1) reduction of Pt-containing ions occurs and the resulting Pt
atoms might form very small clusters, (2) reduction of Pd-containing ions occurs and the resulting
Pd atoms are deposited onto small Pt clusters.   This appears to be the reason why Pt/Pd bimetallic
particles are smaller and are formed with a uniform shape compared with Pt and Pd monometallic

133

particles. A similar phenomenon was previously reported for Pt/Pd bimetallic nanoparticles prepared by mild reduction in solution [12] .

By UV-VIS spectroscopy, sonication of the S(4:1) mixture behaved similarly to S(1:1), however some larger agglomerates (ca. 10 nm) were observed in electron micrographs. In the case of S(1:4), rapid increase of the base line was observed, which is similar to the spectral change for S(0:1). Two particle sizes ranges, of average diameters ca. 3 nm and ca. 4 nm were observed in electron micrographs of S(1:4). Results suggest the coexistence of Pt/Pd bimetallic particles and Pd monometallic particles, which is different from the mild reduction of a Pt and Pd mixture [12]. Since the reduction by ultrasound proceeds faster than the mild reduction method, excess Pd atoms might form pure Pd particles.

CONCLUSIONS

Pt/Pd bimetallic nanoparticles were prepared by the ultrasonic irradiation of aqueous solutions containing Pt and Pd ions (as their chlorometallate anions $PdCl_4^{2-}$ and $PtCl_6^{2-}$). Although the reduction of the Pt solutions proceeds slower than that of Pd , the formation of Pt atoms and/or small Pt clusters occurs prior to the reduction of Pd ions in the solution containing both ions. Electron microscopy revealed that the Pt/Pd bimetallic particles have a similar size to the Pt monometallic particles and have a similar shape to the Pd monometallic particles. Catalysis using the Pt/Pd bimetallic nanoparticles is in progress.

REFERENCES

1.   N. Toshima, K. Kushihashi, T. Yonezawa and H. Hirai, Chem. Lett., 1769(1989).
2.   G. Schmid, Materials Chemistry and Physics, **29**, 133(1991).
3.   A.A. Schmidt and R. Anton, Surf. Sci., **322,**, 307(1995).
4.   G. Schmid, H. West, H. Mehles and A. Lehnert, Inorg. Chem., **36**, 891(1997).
5.   E.B. Flint and K. Suslick, Science, **253**, 1397(1991).
6.   K. Suslick, S.-B. Choe, A.A. Cichowlas and M.W. Grinstaff, Nature, **353**, 414(1991).
7.   Y. Nagata, Y. Watanabe, S. Fujita, T. Dohmaru and S. Taniguchi, J. Chem. Soc., Chem. Commun., 1620(1992).
8.   S.A. Yeung, R. Hobson, S. Biggs and F. Grieser, J. Chem. Soc., Chem. Commun, 378(1993).
9.   Y. Mizukoshi, K. Okitsu, Y. Maeda, T. Yamamoto, R. Oshima and Y. Nagata, J. Phys. Chem. B, **101**, 7033(1997).
10. R. Hiller, S.J. Putterman and B.P. Barber, Phys. Rev. Lett., **69**, 1182(1992).
11. E.B. Flint and K. Suslick, J. Phys. Chem., **95**, 1484(1991).
12. N. Toshima, J. Macromol. Sci. Chem., **A27**, 1225(1990).

Part III

Carbon-Based Catalysts

CATALYTIC PROPERTIES OF FULLERENE MATERIALS

R. MALHOTRA*, A.S. HIRSCHON*, D.F. MCMILLEN*, W.L. BELL*
*SRI International, Menlo Park, CA 94025, ripu@sri.com
**TDA Research, Wheat Ridge, CO 80033, USA

ABSTRACT

Fullerenes were found to catalyze coupling and transalkylation reactions of mesitylene, engage in transfer hydrogenations with dihydroaromatics, and cleave strong bonds such as those in diarylmethanes. In all of these reactions, fullerenes show a marked ability to accept and to transfer hydrogen atoms. The key structural feature that endows fullerenes with many of its characteristics is the presence of a pentagon surrounded by hexagons. We suspected that fullerene soot, unlike graphitic carbon, contained pentagons in a hexagonal lattice, and that these sites imparted the soot with the desired chemical attributes of strong electrophilic nature and an ability to stabilize radicals. In subsequent studies, we have shown fullerene soot to be very effective in catalyzing various H-transfer reactions, including the conversion of methane into higher hydrocarbons. When compared with other carbons, such as activated carbons and acetylene black, the fullerene soot is much more reactive for oligomerization and hydrodealkylation of alkylbenzenes. Because this activity remains, even in chemically extracted and partially oxidized soot, the observed catalysis is not a result of residual soluble fullerenes.

INTRODUCTION

Fullerene Materials

The discovery of fullerenes in the 1980s and the development of a method for their bulk production in 1990 provides the basis for the development of completely new carbon materials. Fullerenes are all-carbon cage molecules, the most celebrated among which is the soccer-ball shaped C_{60}. While such structures were hypothesized in 1970, it was not until 1985 that the first experimental evidence of their existence was found. The structure of fullerenes can be readily understood by considering the structure of a graphene sheet, which consists of carbon atoms in a flat hexagonal lattice. In this lattice, the valencies of the atoms in the middle are fully satisfied, but those at the edges are not, giving rise to slight destabilization. If the lattice is very large, the overall destabilization on a per carbon basis is small. Moreover, many flat lattices can assemble in layers and gain stability due to interplanar van der Waals interactions. However, if the lattice consists of relatively few atoms, destabilization is significant and the system seeks to minimize its internal energy by folding onto itself. Incorporating pentagons in the lattice is one way to achieve folding without having "unsatisfied" valences. Each pentagon introduces a curvature of 60° ($\pi/3$ solid angle). It takes twelve pentagons in a hexagonal lattice for complete closure (4π solid angle). One can also arrive at the same conclusion from Euler's theorem on the relationship between the number of edges, faces and vertices in a closed polyhedron. Thus, C_{60} consists of 20 hexagons and 12 pentagons, C_{70} consists of 25 hexagons and 12 pentagons and C_{84} consists of 32 hexagons and 12 pentagons. The relative location of the pentagons in these strucures creates different shapes, and consequently there can be a wide range of structures for each of these clusters, each with its unique properties.

The range of structures extends beyond that of the single-shell fullerene structures and encompasses large multishelled structures, such as nanotubes and nanopolyhedral particles. The common structural element in all of these materials is the pentagon surrounded by hexagons. The curvature caused by the presence of a pentagon endows the structure with certain chemical properties, including the ability to act as a catalyst for hydrogen-transfer reactions. This structural feature is also likely to be present in the fullerene soot, which comrises about 80% of the material formed during the arc discharge method of producing fullerenes. We surmise that the fullerene

Mat. Res. Soc. Symp. Proc. Vol. 497 © 1998 Materials Research Society soot was formed by those clusters of carbon that did not congeal into self-contained closed-shell species, and these structures continued to grow into large macromolecular structures. Neverthelesss, in this process they too acquired the essential stuctural motif of a pentagon surrounded by hexagons. Hence, we expected the fullerene soot, like the soluble fullerenes, to be a good catalyst for the H-transfer reactions

In this paper we describe some of the results that illustrate the effect of fullerenes and fullerene soot on hydrodealkylation and hydrogenation of aromatic systems, and then briefly summarize our findings on the fullerene soot catalyzed conversion of methane into higher hydrocarbons.

EXPERIMENTAL

Initial stdies on hydrodealkylation were conducted by placing the substrate, fullerene and any solvent in a in fused-silica ampoules. The ampoule was evacuated, sealed, then heated by plunging it into a in a molten-salt bath. After the appropriate period the ampoule was removed, cooled, broken into sections and extracted with a solvent. The reaction products were analyzed by GC or GC/MS.

For reactions conducted under hydrogen pressure, we used a 50-cc shaking autoclave ("Shakerclave," Fluitron Corp.), modified with an aluminum block insert to accept 3.3-cc micro reactors connected to the gas pressure source with 0.040-in id tubing with very small dead volume. The reactors are completely enclosed by the aluminum block, thus eliminating the temperature uncertainties often associated with fluidized sandbaths. The reactors can be operated with fritted metal filters if there is concern about plugging of the connecting tubing. The substrates tested were incorporated into a mixture (typically containing mesitylene, 1,3-dimtheylnaphthalene, 1-benzylnaphthalene, 1,1'-dimethylbiphenyl, 9,10-dimethylanthracene, and 1-methylpyrene). The test mixture is liquid at room temperature and can conveniently be introduced in small quantities via micropipet, along with various H-donor or non-donor additives. The total quantity of reaction mixture is typically about 0.5 g, leaving about 85% head-space for H_2 pressurization.

RESULTS AND DISCUSSION

H-transfer reactions

During our investigations on the extraction of fullerenes from soot using various solvents, we discovered that the fullerenes can engage in a range of H-transfer reactions.[1] For example, when heated under reflux (162°C) in the presence of C_{60}, mesitylene produces dimers and trimers, as well as transalkylation products under relatively mild conditions (Fig. 1).

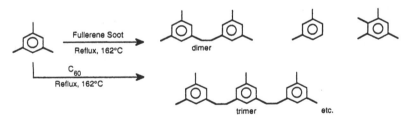

Fig. 1. Fullerenes catalyze coupling and transalkylation reactions.

In view of this marked facility with which fullerenes catalyzed the transfer of hydrogen between mesitylene molecules, we studied the catalytic effect of C_{60} on the cleavage of strong

C_{aryl}-C_{alkyl} bonds, which are also susceptible to cleavage following the H-transfer. We found that the addition of C_{60} increased the cleavage rate of 1,2'-dinaphthylmethane in each of the three hydroaromatic/aromatic solvent systems that we studied (tetralin/napthalene; 9,10-dihydro-anthracene/anthracene, and 9,10-dihydrophenanthrene/phenanthrene). The selectivity of cleavage at the 1-position (which leads to the formation of 2-methylnaphthalene) was also increased in each case. Higher reactivity in the face of higher selectivity indicates that the increase in the number of H-carriers more than compensates for the increased selectivity of those generated from C_{60}. Moreover, since the background reaction (i.e., without C_{60}) contributed to about half of the net cleavage, the selectivity of the C_{60}-catalyzed path is even greater than that observed for the total reaction.

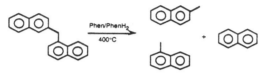

Fig. 2. Fullerenes catalyze cleavage of strong C_{aryl}-C_{alkyl} bonds.

In these reactions, we also observed the hydrogenation of fullerenes by the hydroaromatic. Prolonged heating, however, led to the dehydrogenation of the partially hydrogenated fullerene to give back the fullerene and (presumably) hydrogen. Thus we can say that the fullerene catalyzed the dehydrogenation of the hydroaromatic. Fullerenes such as C_{60} were also found to undergo hydrogenation reactions with molecular H_2 at 400°C in the absence of other catalysts giving a broad range of range of hydrogenated fullerenes, $C_{60}H_x$ (x = 2 to 40).

Although the initial hydrodealkylation study was conducted with C_{60}, we suspected that the fullerene soot formed during the arc process might also serve as a good catalyst. This soot probably results from carbon clusters that did not completely close on themselves, hence went on to combine with other clusters with dangling bonds. If this is true, the fullerene soot will contain pentagons in a hexagonal lattice, and these sites will impart the soot with the desired chemical attributes of a strong electrophilic nature and an ability to stabilize radicals—attributes that make it an effective catalyst for hydrogen-transfer reactions.

We conducted a series of tests with several model substrates the probe the efficacy of fullerene soot and other carbons to promote hydrodealkylations. Fig. 3 compares the observed first-order rate constant for dealkylation of dimethylnanphthalene in tetralin at 400°C in the absence and presence of various carbons. In the absence of added carbon, the rate constant is on the order of 1.0×10^{-6} s^{-1}; it increases to about 5×10^{-6} in the presence of fullerene soot. Partial oxidation of the soot increses the surface area from 125 m^2/g to roughly 700 m^2/g, but it does not appear to augment the soot-catalyzed dealkylation. Acetylene black and Norit carbons also increase the rate of dealkylation but not as much as the fullerene soot.

Similar reactions with methylpyrene show somewhat different, but not unexpected, results. Addition of hydrogen atom to a pyrene nucleus is significantly more exothermic than to the naphthalene nucleus (38 kcal/ml vs. 30 kcal/mol). Thus, hydrodealkylation rates are considerably higher with methylpyrene, and they are somewhat less subject to catalysis. As can be seen in Fig. 4, with the exception of the extracted but un-oxidized soot, the increase is a factor of 2-3, rather than 4-6, as was the case for the less reactive dimethylnaphthalene.

139

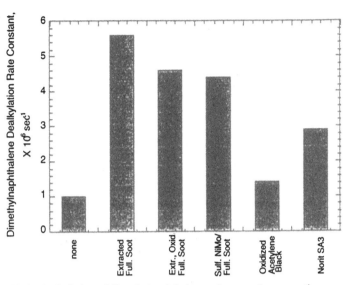

Figure 3. Hydrodealkylation of dimethylnaphthalene, a less reactive aromatic, promoted by various carbons.

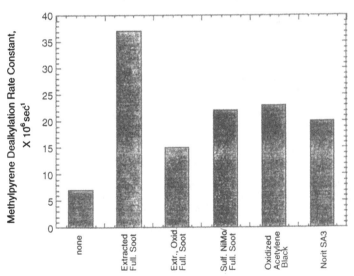

Fig. 4. Hydrodealkylation of methylpyrene, a reactive aromatic, promoted by various carbons.

The above studies were conducted in the absence of any hydrogen pressure, and we wondered if the catalytic effect of fullerene soot observed in the case where a hydroaromatic, such as tetralin, was the source of hydrogen, would also be seen in the presence of excess hydrogen pressure. Accordingly, we conducted several runs in pressurized microreactors. The results for the hydrodealkylation of several alkylaromatics, shown in Fig. 5, clearly demonstrate that fullerene soots also enhance the rates of hydrodealkylation in the presence of hydrogen pressure.

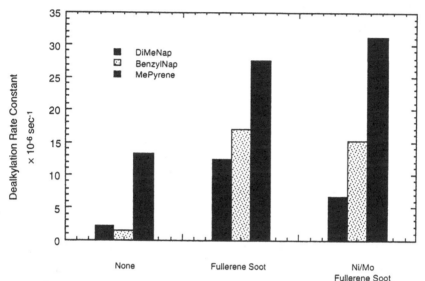

Fig. 5. Hydrodealkylation of alkylaromatics promoted by fullerene soot and Ni/Mo-loaded fullerene soot under hydrogen pressure.

Methane Activation

The direct hydrogenation of C_{60} by molecular H_2 shows the ability of fullerenes to activate bonds as strong as the 104-kcal/mol H-H bond. This bond has roughly the same strength as that of the C-H bond in methane (105 kcal/mol). Hence, fullerenes appear to possess many of the attributes necessary for methane activation, and we postulated that fullerenes may provide a surface for the alkyl radicals to chemisorb, thereby moderating their reactivity.[2] We used fullerene soot in the study of methane conversion, because the common soluble fullerenes, such as C_{60}, C_{70}, and C_{84}, may evaporate during the reaction.

The methane activation experiments were conducted in a flow reactor with methane at atmospheric pressures. The catalyst was supported on a fritted disk in a quartz reactor. The exit of the reactor was fitted with a quenching zone through which cooling gases could be passed. The reactor system was placed in a dual furnace system where the methane was first heated in a preheating furnace (at 600°C) and then passed into a high temperature furnace (containing the catalyst bed). The high temperature furnace contains a short heating zone (4") which leads to a cool quenching zone to minimize any possible thermal reactions after conversion. The gaseous products were analyzed by a Carle 500 gas chromatograph equipped with a thermal conductivity detector and hydrogen transfer tube for hydrogen analysis. Argon was used in the methane gas stream as an internal standard.

Fullerene soot was found to lower the temperature-threshold for methane conversion by 250°C relative to the thermal case, and by about 150°C over other forms of carbon (acetylene black and Norit-A) as shown in Fig. 6.[2] No condensible products were formed in the presence of fullerene soot or other carbons. Neither an increase in surface area by partial oxidation with CO_2 nor extraction of the soluble fullerenes appears to alter the reactivity of the fullerene soot. The conversion of methane with soot decreased with the flow rate while the C_2 yield increased accordingly. These results are consistent with the fullerene soot, not only being active for methane conversion, but also being active for subsequent reactions with products such as ethane and ethylene leading to lower C_2 yields and coke at lower flow rates.

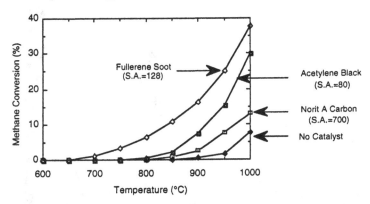

Fig. 6. Conversion of methane over various carbons (CH_4 flow rate 100 cc/min).

Evidently, the surface-bound methyl radicals were losing hydrogen faster than they could combine with each other. Thus, if we could find metals that would inhibit the loss of hydrogen from the fullerene surface, we might be able to provide a greater opportunity for the methyl radicals on the surface to combine with each or otherwise engage in bimolecular reactions. Accordingly, we conducted experiments in which the fullerene soot was doped with various metals. We found that doping the fullerene soot with potassium or managanese was particularly effective in increasing the selectivity to higher hydrocarbons from about 25% to greater than 70% at identical conditions where the conversion levels were approximately 5%. Although neither the catalyst nor the process was optimized, these results provide direction for future work with fullerene based catalysts in order to improve the methane conversion process.[3]

2.3 Fullerenes as catalyst supports

The effect of C_{60} as a ligand on the Rh-catalyzed hydroformylation was studied by Claridge et al., who found notable changes in the turnover frequency and n/iso-ratio for the reaction with propene, but not with ethene.[4] Fullerene materials have also been used as supports for other metal catalysts. In the case of catalitically grown carbon fibers, Baker has suggested that because most of the surface this this form of carbon presents to the metal crystal is the edge surface, as opposed to the basal plane that graphitic forms present, the activity of the metal crystal will be modified.[5] Planeix and coworkers studied the reduction of cinnamaldehyde by ruthenium supported on alumina, carbon, or nanotubes. They found that the regioselectivity to cinnamyl alcohol was between 20% and 30% when alumina was the support material. The selectivity increased somewhat to about 35% with activated carbon, however, it was greater than 90% when nanotubes were used as the support material.[6]

142

CONCLUSION

There is a wide range of fullerene materials that is now available in significant quantities. These materials share the structural feature of a pentagon surrounded by hexagons, which endows them with electrophilicity and an ability to stabilize radicals. Even fullerene soot, which is formed as a byproduct during the production of fullerenes, has some of these properties. We have shown that fullerene materials, including fullerene soot, are excellent catalysts for H-transfer reactions such as hydrogenation and hydrodealkylations. They are also highly effective in promoting the conversion of methane into higher hydrocarbons. The selectivity of C_2 and higher species can be substantially improved by doping the fullerene soot with metals such as potassium or manganese.

ACKNOWLEDGMENTS

Financial support for the initial phase of this work was provided by the New Energy and Industrial Development Organization (NEDO) of the Ministry of International Trade and Indusry (MITI), Japan. Subsequent studies were conducted under the sponsorship of the US Dept. of Energy.

REFERENCES

1. R. Malhotra, D. F. McMillen, D. S. Tse, D. C. Lorents. R. S. Ruoff and D. M. Keegan, *Energy Fuels* **7**, 685 (1993).
2. A. S. Hirschon, H.-J. Wu, R. B. Wilson and R. Malhotra, *J. Phys. Chem.* **99**, 17483 (1995).
3. A. S. Hirschon, Y. Du, H.-J. Wu, R. B. Wilson and R. Malhotra, *Res. Chem. Intermed.* **23**, 675 (1997).
4. J. B. Claridge, et al., *J. Mol. Catal.* **89**, 113 (1994).
5. R. T. Baker, personal communication.
6. J. M. Planeix, et al., *J. Am. Chem. Soc.* **116**, 7935 (1994).

THE IMPACT OF THE GRAPHITE NANOFIBER STRUCTURE ON THE BEHAVIOR OF SUPPORTED NICKEL

C. PARK and R. T. K. BAKER
Department of Chemistry, Northeastern University, Boston, MA 02115

ABSTRACT

In the current investigation we have used the hydrogenation of ethylene and crotonaldehyde as probe reactions in an attempt to follow any changes in catalytic behavior induced by supporting nickel on different types of graphite nanofiber support materials. The hydrogenation of the α,β-unsaturated aldehyde to the desired product, crotyl alcohol, is a particularly difficult task since there is a strong tendency to hydrogenate both the C=C and C=O in the reactant molecule. This study is designed to compare the catalytic behavior of the metal particles when dispersed on three types of nanofibers, where the orientation of the graphite platelets within the structures is significantly different in each case. The metal crystallites are located in such a manner that the majority of particles are in direct contact with graphite edge regions. For comparison purposes, the same set of hydrogenation reactions were carried out under similar conditions over γ-Al$_2$O$_3$ supported nickel particles.

INTRODUCTION

The electrical conductive properties of graphite offer some interesting possibilities when the material is used as a catalyst support medium for small metal particles. Unfortunately, the relatively low surface area, approximately 10 m^2/g, of conventional forms of graphite creates a major drawback for such an application. This limitation can be readily overcome if one chooses to use graphite nanofibers for this purpose, since it is possible to tailor the orientation of the graphite platelet stacks that constitute the material and generate structures that exhibit high surface areas of between 100 to 700 m^2/g [1]. Previous studies by Rodriguez and coworkers [2] showed that Fe-Cu particles dispersed on graphite nanofibers exhibited a much higher catalytic activity for the conversion of hydrocarbons than when the same concentration of the bimetallic supported on traditional carriers such as activated carbon or γ-alumina. A subsequent investigation by Chambers and coworkers [3] established that the performance of nickel particles supported on a variety of carbonaceous materials was extremely sensitive to the degree of crystalline perfection of the substrate; the highest activity for the hydrogenation of 1-butene and 1,3-butadiene at 100°C being found for samples in which the metal particles were supported on carbon nanofibers possessing a high graphitic content. These results were in accord with those reported by other workers [4] who claimed that for the hydrogenation of selected hydrocarbons, graphite supported palladium particles exhibited the highest activity and selectivity compared to that found when the metal was dispersed on less ordered carbonaceous solids. Gallezot and coworkers [5] showed that significant differences in both activity and selectivity were achieved for the hydrogenation of cinnamaldehyde when platinum particles were supported on graphite compared to that found when the metal was dispersed on high surface area amorphous carbon.

In the current study we have used the hydrogenation of ethylene and crotonaldehyde as probe reactions in an attempt to assess any modifications in catalytic behavior induced by supporting nickel on three types of nanofibers, where there were significant differences in the arrangements of the graphite platelets [6]. It was speculated that the atomic arrangement of the surfaces of nickel particles that nucleated on these different graphite edges would be dictated to a large degree by the interaction with the atoms in nanofiber supports. Under such circumstances one might reasonably expect that different crystallographic faces of nickel would be exposed to the reactant gas depending on which type of nanofiber structure was used as the supporting medium. The success of this approach could open up the possibility of a tailoring the morphological characteristics of metal particles in such a fashion so to achieve a desired catalytic performance.

The selective hydrogenation of α,β-unsaturated aldehydes to give the corresponding unsaturated alcohol presents a challenge both from the point of view of thermodynamic and

145

kinetic considerations [7]. Existing catalysts are based on alumina supported noble metals and do not offer a high selectivity towards the desired product [8]. A theoretical study concludes that both the nature of the metal and the type of exposed crystal face are critical factors in determining the performance of the catalyst system for this reaction [9]. English and coworkers [10] demonstrated that the hydrogenation of crotonaldehyde over Pt/SiO_2 and Pt/TiO_2 catalysts was strongly influenced by the size of the metal particles and the promotional effects of surface oxides. They reported that the selectivity to the unsaturated alcohol in the gas phase hydrogenation reaction increased with increasing metal particle size.

EXPERIMENTAL

Materials

    Figure 1 is a schematic presentation of the three types of graphite nanofibers used in this work and were grown according to the procedures outlined previously [6]. The nanofibers are classified as possessing a "platelet" structure in which the graphene sheets are oriented perpendicular to the fiber axis, (Figure 1a); a "spiral" form where the platelets are also aligned perpendicular to the fiber axis (Figure 1b); and a "ribbon-like" structure where the platelets are arranged parallel to the fiber axis (Figure 1c). Prior to use as support media, the catalyst particles responsible for creating these nanofiber structures from the decomposition of various carbon-containing gases, were removed by dissolution in 1M hydrochloric acid. This step was followed by a thorough wash in deionized water before being dried overnight in air at 110°C.

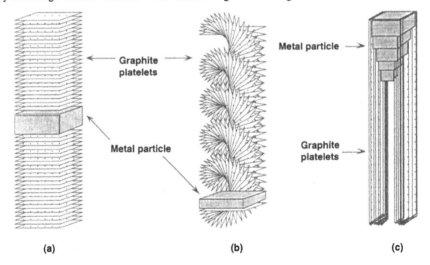

(a)   (b)   (c)

FIGURE 1 Schematic rendition of various graphite nanofiber conformations

    The 5 wt.% supported nickel catalysts used in this study were prepared by a standard incipient wetness technique. Each support medium was impregnated with a ethanol solution containing the appropriate amount of the precursor salt necessary to achieve the desired metal loading. After drying overnight at 110°C, the impregnated materials were calcined in air at 250°C for 4 hours and then reduced at 350°C in a 10% H_2/He mixture for 20 hours. The reduced catalyst samples were cooled in He and finally, passivated in 2%O_2/He for one hour at room temperature before being removed from the reactor. In the case of the γ-alumina supported nickel sample this protocol was modified, the reduction step in H_2 at 350°C was increased for periods up to 60 hours in order to ensure that complete conversion of nickel to the metallic state was being realized.

146

The gases used in this work, helium (99.999%), hydrogen (99.999%), carbon monoxide (99.99%), ethylene (99.95%), were obtained from MG Industries and used without further purification. Reagent grade nickel nitrate [$Ni(NO_3)_2.6H_2O$] and crotonaldehyde (99.9%) were obtained from Fisher Scientific and Aldrich Chemical Co., respectively. The γ-alumina was supplied by Degussa Corporation and was used in the as-received condition.

Apparatus and Procedures

Catalyst studies were performed in a vertical quartz reactor, fitted with a quartz frit at the mid-point of the tube, and this assembly was located within a split furnace. The gas flow to the reactor was regulated by MKS mass flow controllers allowing for a constant reactant feed to be delivered to the system. Catalyst samples (100 mg) were positioned on the quartz frit and the tube aligned in such a manner that this region was in the hot zone of the furnace. After reduction in a 10%H_2/He mixture for 2 hours at 400°C the system was cooled to the desired reaction temperature and the reactant gas admitted to the catalyst for periods of up to 3 hours. In the case of crotonaldehyde, when the catalyst system reached the desired temperature the gas flow of the 10%H_2/He mixture was diverted through a saturator containing the liquid reactant maintained at a constant temperature of 20°C. The reactions were conducted under differential conditions in order to avoid mass transfer limitations. The progress of a given reaction was monitored as a function of time by gas chromatography analysis of the inlet and outlet streams at regular intervals.

The supported nickel catalysts were characterized by a variety of techniques, including high resolution transmission electron microscopy (HRTEM), X-ray diffraction, temperature programmed oxidation (TPO) and BET surface area measurements. In addition, the catalyst samples were also examined by HRTEM to ascertain the existence of any changes in either the support structure or morphology of the metal particles induced by treatment in the various hydrocarbon/hydrogen mixtures.

RESULTS AND DISCUSSION

Characterization Studies

A combination of CO_2 temperature programmed oxidation experiments and X-ray diffraction measurements established that all the nanofibers used in this work were highly graphitic in nature with an extremely low amorphous carbon content. TEM studies revealed that the nanofibers were between 100 to 150 nm in width and varied from 5 to 50 µm in length. Examination of the freshly prepared nanofiber supported nickel particles showed that in all cases, the metal particles tended to be located along the edge sites of each type of nanofiber structure. From detailed inspection it was possible to establish that the metal particles were on average about 7.0 nm in width and adopted a well-defined hexagonal outline. The morphological characteristics displayed by these particles were consistent with the existence of a strong metal-support interaction. In sharp contrast, it was extremely difficult to determine the shape of the particles on the γ-alumina support since the average width was only 1.4 nm.

Flow Reactor Studies

(a) Catalytic hydrogenation of ethylene by supported nickel particles

It is apparent from the results shown in Table 1 that the ability of nickel to catalyze the hydrogenation of ethylene at 100°C does not appear to exhibit any dramatic variations when the metal is supported on the different media. An overall high conversion of the olefin, primarily to ethane and to a lesser extent, solid carbon was achieved when a C_2H_4/H_2 (1:1) mixture was passed over each supported nickel catalyst system at 100°C. Clearly in this reaction metal particle size does not play a dominant role, as the introduction of nickel onto nanofibers possessing a tubular structure was found to exhibit a marginally higher activity for this reaction, than that of

147

the other systems despite the fact that the particle sizes were about six times larger on this material than those on the γ-alumina.

TABLE 1. Percent conversion of ethylene to gaseous and solid products over various supported nickel catalysts after 90 mins at 100°C.

Catalyst Support	% Conversion C_2H_4 to Products		
	CH_4	C_2H_6	C_s
Ribbon GNG	0.02	63.69	8.60
Spiral GNF	0.04	71.64	11.61
Platelet GNF	0.01	53.16	7.37
γ-Alumina	0.03	68.78	10.58

An interesting feature was observed with a catalyst comprised of nickel supported on the "platelet" type of graphite nanofibers and this was the finding that at 100°C there was an induction period of about 35 mins required for this system to achieve an overall activity comparable to that displayed by the other supported metal catalysts. This reactivity sequence shown in Table 1 was found to persist when the temperature was increased up to a maximum level of 200°C.

One might contend that the observed behavior of nickel/γ-alumina is subject to artificially low values resulting from incomplete reduction of the metal when dispersed on the oxide carrier. There are a number of factors that refute this possibility; hydrogen chemisorption measurements gave a particle size that was very close to that determined from TEM analyses. The 5 wt.% nickel/alumina catalyst exhibited a relatively high activity for the hydrogenation of ethylene and portions of the same sample were subsequently used for hydrogenation reactions of other more complex organic molecules. Furthermore, the catalytic behavior of alumina supported nickel particles did not exhibit any discernible differences as a result of increasing the time of reduction in the preparative step from 20 to 60 hours.

(b) Catalytic hydrogenation of crotonaldehyde by supported nickel particles

In this series of experiments 5 wt.% nickel was supported on two types of graphite nanofibers, platelet and the ribbon forms, and the performance of these systems was compared to that exhibited by the same metal loading dispersed on γ-alumina. Inspection of the data shown in Table 2 indicates that the graphite nanofiber supported nickel particles exhibit a significantly higher activity for the conversion of crotonaldehyde than that displayed by the Ni/alumina system. It is interesting to find that the orientation of graphite platelets in the nanofiber supports also exerts an impact not only on the overall activity, but also on the selectivity pattern. This latter feature can be readily observed from Figure 2, which shows the percent selectivity towards the formation of the desired product in this reaction, crotyl alcohol. In both regards it is evident that the superior performance is achieved from dispersing the metal crystallites on the platelet type of nanofibers.

It is probable that when nickel is supported on the nanofibers, the crystallites adopt morphological characteristics that are quite different to those that exist when the metal is dispersed on more traditional materials. The surfaces exposed by the platelet nanofibers consist of two similar sized-edges that are oriented to an equal degree either in the "armchair" or "zig-zag" configurations. While it is not possible to determine whether nickel particles will preferentially nucleate at one or other of these sites it is clear that the atomic arrangements of the metal atoms will be quite different at these locations. As a consequence, it is expected that the interactions with the respective edge regions will dictate the catalytic behavior of the nickel crystallites. It would appear that under these circumstances that the atomic arrangement of the metal particles supported on graphite nanofibers favors the adsorption of crotonaldehyde in such a manner that hydrogenation of the C=O bond is a facile process, whereas activation of the C=C bond in the molecule is prevented.

TABLE 2. Percent conversion of crotonaldehyde over various supported nickel catalysts as a function of reaction temperature.

Temperature (°C)	Ni/γ-Alumina	Ni/Ribbon GNF	Ni/Platelet GNF
75	3.31	19.28	73.92
100	25.60	55.60	89.27
125	42.00	79.78	98.23
150	82.60	99.44	99.89

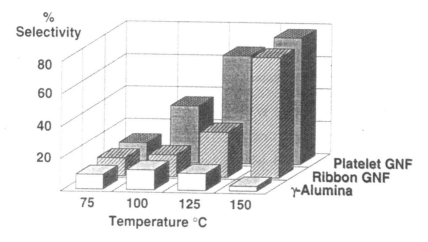

FIGURE 2. Selectivity towards crotyl alcohol formation from the hydrogenation of crotonaldehyde over various supported nickel catalysts as a function of reaction temperature

CONCLUSIONS

The morphological characteristics of small metal particles and the role played by the supporting medium on such features is an area of extreme importance in heterogeneous catalysis. The results of the present investigation demonstrate that while the use of highly graphitic nanofibers exhibits little benefit for the behavior of supported nickel particles in the hydrogenation of C_2H_4, these unusual carbonaceous materials exert a tremendous impact on the performance of the metal for the hydrogenation of crotonaldehyde to crotyl alcohol. This enhancement in both activity and selectivity is attributed to the fact that for the most part nickel crystallites are located on the edge sites of the carbon nanofibers and as a consequence, the arrangement of the metal atoms will be governed to a large degree by the interaction with the carbon atoms in these regions. Under such circumstances one might reasonably expect that different crystallographic faces of nickel will be exposed to the reactant gas compared to those that are present when the metal is dispersed on less ordered materials such as γ-alumina, silica or active carbon.

149

ACKNOWLEDGMENTS

The authors would like to thank Prof. N. M. Rodriguez for stimulating discussions and help with the HRTEM examinations. The work was supported by a grant from the National Science Foundation, Division of Chemical and Transport Systems, No. CTS-9634266.

REFERENCES

1.   N. M. Rodriguez, J. Mater. Res. **8**, 3233 (1993).

2.   N. M . Rodriguez, M. S. Kim and R. T. K. Baker, J. Phys. Chem. **98**, 13108 (1994).

3.   A. Chambers, T. Nemes, N. M. Rodriguez and R. T. K. Baker, submitted to J. Phys. Chem.

4.   I. C. Brownlie, J. R. Fryer and G. Webb, J. Catal. **64**, 263 (1969).

5.   P. Gallezot, D. Richard, and G. Bergeret, in "Novel Materials in Heterogeneous Catalysis", (R. T. K. Baker and L. L. Murrell, Eds.) ACS Symposium Series, **437**, 150 (1990).

6.   N. M. Rodriguez, A. Chambers and R. T. K. Baker, Langmuir **11**, 3862 (1995).

7.   F. Coloma, A. Sepulveda-Escribano and F. Rodriguez-Reinoso, Appl. Catal. **150**, 165 (1997).

8.   M. A. Vannice, J. Mol. Catal. **59**, 165 (1990).

9.   F. Delbecq and P. Sautet, J. Catal. **142**, 21 (1995).

10.   M. English, L. Andeas and A. Johannes, J., Catal. **166**, 25 (1997).

ESR OF PURIFIED CARBON NANOTUBES PRODUCED UNDER DIFFERENT HELIUM PRESSURES

S.P.WONG*[#], HAIYAN ZHANG**, NING KE*, and SHAOQI PENG**
*Department of Electronic Engineering, The Chinese University of Hong Kong, Hong Kong, China
**Department of Physics, Zhongshan University, Guangzhou 510275, China

ABSTRACT

Carbon nanotubes were prepared by the dc arc-discharge method under a controlled helium pressure ranging from 10 to 80 kPa and subsequently purified by oxidation in air. The purified carbon nanotubes were observed by transmission electron microscopy. The room temperature electron spin resonance (ESR) spectra of the purified nanotubes were measured. The variations in the ESR line shape, g-value, linewidth and relative spin density of the purified nanotubes on helium pressure were studied and discussed.

INTRODUCTION

Since the theoretical studies predicted that carbon nanotubes with nanometer-scale diameters would exhibit diversified electronic properties depending on their diameters and degree of helicity [1-4], there has been a great interest in the study of the electronic and magnetic properties of carbon nanotubes. Ebbesen and Ajayan [5] first determined the conductivity of the bulk nanotube materials consisting of tubes of various diameters to be of the order of 10^4 Sm^{-1} and they inferred that some tubes must have a much higher conductivity than this average value. More recent measurements indeed showed that individual nanotubes could be either metallic or non-metallic [6].

On the other hand, electron spin resonance (ESR) measurements have been employed to detect conduction electrons in these nanotube structures and hence to tell whether metallic or very narrow bandgap semiconducting tubes are present [7, 8]. The conduction electrons of metallic and narrow gap semiconducting nanotubes, like those in graphite [9, 10], are expected to lead to detectable conduction electron spin resonance (CESR) signals. Purified carbon nanotubes can be obtained by oxidizing the crude nanotube materials in air at high temperatures so that the carbon nanoparticles in the cathodic deposits can gradually be removed [11]. On ESR study of the purified nanotubes, there are very different results reported in the literature [7, 8]. For example, in one work, it was concluded that purified nanotubes are ESR silent at room temperature [7]. In another work, Kosaka et al. [8] performed an ESR study of carbon nanotube samples at various stages of purification by oxidation and identified the ESR signals which are due to conduction electrons in purified nanotubes.

In this work, we shall present our recent results on room temperature ESR measurements of purified carbon nanotubes produced under different helium pressures. The effects of He pressure (P_{He}) on the ESR spectral line shape, g-value, linewidth and relative spin density of purified nanotubes are studied and discussed.

[#]E-mail: spwong@ee.cuhk.edu.hk

Mat. Res. Soc. Symp. Proc. Vol. 497 © 1998 Materials Research Society

EXPERIMENTAL

The crude nanotubes containing pure nanotubes and nanoparticles were produced by the dc arc-discharge method in helium gas at a controlled pressure ranging from 10 to 80 kPa. The applied voltage was 26V and the electric current was 60A. The crude nanotubes were then oxidized at 770°C in air until about 1% of the initial weight remained. The samples were observed by transmission electron microscopy (TEM). The ESR spectra of the samples were measured using a JOEL JES-FEIX electron spin resonance spectrometer operating at 9.46 GHz with a field modulation frequency of 100 kHz at room temperature. The Mn^{2+} ion was used as a reference marker of the g-values.

RESULTS AND DISCUSSIONS

Typical TEM micrographs of purified nanotubes produced under two different P_{He} values are shown in Fig. 1. It is seen that nanotubes produced at a lower P_{He} pressure is "cleaner" and the tubule radii are generally smaller compared with those of nanotubes produced at a higher P_{He}. The diameters of the nanotubes produced at 11 kPa P_{He} shown in Fig. 1a range approximately from 2 to 35 nm whereas for those produced at 80 kPa P_{He} shown in Fig. 1b range from 10 to 70 nm. According to tight-binding band structure calculations, the band gap for non-metallic tubes was estimated to be related to the tubule radius R_T by [12]

$$E_g = V_{pp}\, r_{cc}/R_T \qquad (1)$$

where V_{pp} is the hopping matrix element between two carbon p orbits, which is equal to 2.77 eV and r_{cc} is the carbon-carbon bond length equal to 0.14 nm. From Eq. (1) and the TEM results, it is inferred that the band gap of nanotubes produced at higher P_{He} would be smaller than those produced at lower P_{He}. It will be helpful to have a rough idea on the magnitude of the band gap energy according to Eq. (1). For example, $E_g = 0.133$ eV for a R_T value of 3 nm. From the observed tubule radius distribution in our samples, a large fraction of the nanotubes is expected to have narrow band gaps or to be semi-metallic. Hence higher conductivity is expected for nanotubes produced at higher P_{He}.

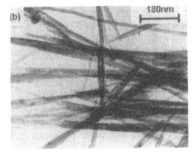

Fig. 1 TEM micrographs of purified nanotubes produced under a He pressure of
(a) 11 kPa and (b) 80 kPa.

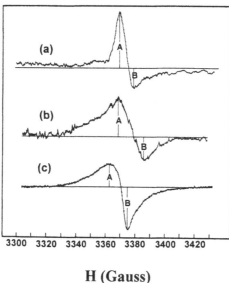

H (Gauss)

Fig. 2 Room temperature ESR spectrum of (a) a crude nanotube sample prepared under a He pressure of 66 kPa; and those of purified nanotube samples prepared under He pressures of (b) 80 kPa and (c) 13 kPa, respectively.

The ESR spectra for a crude nanotube sample and two purified nanotube samples are shown in Fig. 2. The ESR line shape of the crude nanotube sample (spectrum a) and that of the purified sample prepared at higher P_{He} (spectrum b) resemble that of graphite, which exhibits a Dysonian line shape characteristic of CESR in bulk metal samples [9, 10, 13]. This indicates that there are conduction electrons in these nanotube samples, either from some metallic or narrow bandgap semiconducting nanotubes or simply from some graphitic-like structures. The Dysonian-like line shape also suggests that the CESR is affected by the skin depth effect to some extent in these nanotube samples. However, the ESR line shape of the purified nanotube sample prepared at low P_{He} (spectrum c) is quite different from that of graphite. To see how the characteristics of the ESR spectra of the purified nanotube samples vary with P_{He} under which the crude nanotubes were produced, we plot the peak height ratio A/B, the g-value, and the peak-to-peak linewidth, ΔH_{pp}, of these spectra against P_{He} in Fig. 3(a), (b) and (c), respectively. The relative spin densities of these samples are also plotted in Fig. 3(d).

It is seen in Fig. 3(a) that when P_{He} decreases from 80 to 13 kPa, the A/B ratio decreases monotonically from 1.7 to 0.5. For graphite, the A/B ratio has a high value of 3.0 at room temperature [9]. The results on the A/B ratio suggest that purified nanotubes produced at higher P_{He} are more graphite-like than those produced at lower P_{He}. This conclusion is consistent with our earlier discussion on the TEM results and the dependence of band gap on the tubule radius size. Similar conclusion was obtained in a previous study on the thermal properties of nanotubes produced by the arc-discharge method under different P_{He} [14]. It is also noted that the A/B ratio for crude nanotube samples is generally between those of their purified counterparts and

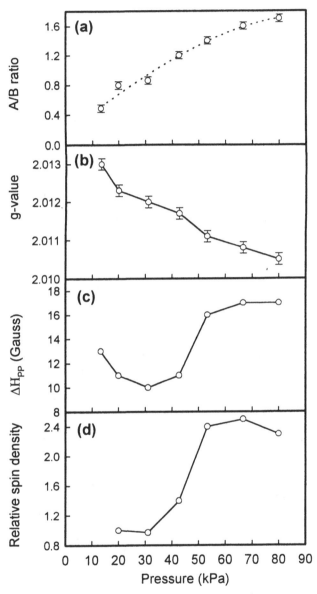

Fig. 3   (a) The peak height ratio A/B; (b) the g-value; (c) the peak-to-peak linewidth ΔH_{pp} of the ESR spectra and (d) the relative spin density of the purified nanotubes versus He pressure under which the crude nanotubes were prepared.

graphite. This is obviously a reflection of the relative "dirtiness" of the crude nanotubes which are likely to have been contaminated to a larger extent by fragments of graphite or some graphitic structures. Note that an ideal Dysonian line shape should have an A/B ratio lying within the range from 2.7 to 19 which is true for CESR signals in thick metal samples [13]. The deviation from the Dysonian line shape can be taken as an indicator of the relative unimportance of the skin depth effect in these samples. A larger deviation means a smaller skin depth effect. We believe that the variation of the A/B ratio with P_{He} is associated with the tubule radius size variation as revealed by the TEM results in this sense.

As shown in Fig. 3(b), the g-value increases monotonically from 2.0105 to 2.013 when P_{He} decreases from 80 to 13 kPa. The g-value of nanotubes was shown to vary during the oxidation purification process and the g-value of the CESR peak for their un-annealed purified nanotube sample was reported to be 2.012 at 296K in ref. [8]. After annealing at 2850°C in an evacuated furnace, the g-value was found to have shifted to 2.0022. Other g-values of purified nanotubes have also been reported in the literature depending on the sample preparation method and the preparation conditions. This should not be surprised as a nanotube sample in fact contains nanotubes of various types, metallic or non-metallic, in various quantities. The CESR signal may have also been contaminated by signals from other defect centers. This work is an example showing how the g-value, and hence also the microstructure, can vary systematically with one particular process parameter, the He pressure. An understanding of the correlation between the g-value and the microstructural, electronic and magnetic properties of the nanotubes is absent at the moment and requires further investigation.

In Fig. 3(c) the peak-to-peak linewidth shows an unexpected minimum at a He pressure of 30 kPa. A smaller linewidth usually corresponds to a less disordered state, a weaker interaction or fewer interacting centers. Does this minimum really associate with a better quality of the nanotubes in this sense? For example, an increase in the ESR linewidth may result from interactions between some defect centers and the conduction electrons. It requires further investigation, for example, with temperature variable ESR measurements, and/or with other characterization techniques to answer this question. In Fig. 3(d), the relative spin density sees an abrupt decrease of more than two fold with decreasing P_{He} in the range from 50 kPa to 30 kPa. The higher spin density at higher P_{He} should again be associated with the more graphite-like nature of the samples produced at higher P_{He}, i.e. higher conductivity and hence higher conduction electron concentration which would lead to a stronger CESR signal. However, whether the abrupt decrease with decreasing P_{He} is associated with some kind of transition in the microstructures or electronic structures of the nanotubes is another question of interest. It is further noticed that in the same range of P_{He}, ΔH_{pp} also sees an abrupt decrease with decreasing P_{He}. Whether this is simply a coincidence or this is really associated with some kind of transition again requires further investigation. With a better understanding of the correlation between these ESR parameters and the microstructures and electronic structures of nanotubes, the ESR parameters could probably serve as indicators of the quality of the nanotubes and could hence shed light on the search for means to produce nanotubes of controlled properties and structures.

CONCLUSIONS

In summary, we have studied the effect of He pressure under which the nanotubes are produced in the arc-discharge on ESR spectra of purified nanotures. The variation of the ESR line shape, the g-value, the linewidth, and the relative spin density with He pressure have been discussed in conjunction with the TEM results on the tubule radius size of the nanotubes. Some interesting features in the He pressure dependence of the ESR parameters are noted. However, an understanding on the correlation between the ESR parameters and the microstructures and electronic structures of the nanotube samples is absent and requires further investigation. Such an understanding would certainly lead to important contribution to the advancement of nanotube science and technology.

ACKNOWLEDGMENTS

This work is supported in part by Guangdong Provincial Natural Science Foundation of China and the Foundation of Key Scientific Researches by the Higher Education Bureau of Guangdong Province, China.

REFERENCES

[1]   J.W. Mintmire, B.I. Bunlap and C.T. White, Phys. Rev. Lett. 68 (1992) 631.
[2]   C.T. White, D.H. Robertson and J.W. Mintmire, Phys. Rev. B47 (1993) 5485.
[3]   N. Hamada, S. Sawada and A. Oshiyamu, Phys. Rev. Lett. 68 (1992) 1579.
[4]   R. Saito, G. Dresseihaus and M.S. Dresseihaus, J. Appl. Phys.73 (1993) 494.
[5]   T.W. Ebbesen and P.M. Ajayan, Nature 358 (1992) 220.
[6]   T.W. Ebbesen, H.J. Lezec, H. Hiura, J.W. Bennett, H.F. Ghaemi, T. Thio, Nature 382 (1996) 54.
[7]   K. Tanaka, T. Sato, T. Yamabe, K. Okahara, K. Uchida, M. Yumura, H. Niino, S. Ohshima, Y. Kuriki, K. Yase and F. Ikazaki, Chem. Phys. Lett. 223 (1994) 65.
[8]   M. Kosaka, T.W. Ebbesen, H. Hiura and K. Tanigaki, Chem. Phys. Lett. 225 (1994) 161; Chem. Phys. Lett. 233 (1995) 47.
[9]   G. Wagoner, Phys. Rev. 118 (1960) 647.
[10] L.S. Singer and G. Wagoner, J. Chem. Phys. 37 (1962) 1812.
[11] T.W. Ebbesen, P.M. Ajayan, H. Hiura and K. Tanigaki, Nature 367 (1994) 519.
[12] J.W. Mintmire, D.H. Robertson and C.T. White, J. Phys. Chem. Solids 54 (1993) 1835.
[13] G. Feher and A.F. Kip, Phys. Rev. 98 (1955) 337.
[14] H. Zhang, D. Wang, X. Xue, B. Chen and S. Peng, J. Phys. D: Appl. Phys. 30 (1997) L1.

REVERSIBLE HYDROGEN UPTAKE IN CARBON-BASED MATERIALS

S. D. M. Brown*, G. Dresselhaus†, and M. S. Dresselhaus*,**
*Department of Physics, Massachusetts Institute of Technology, Cambridge, MA 02139
†Francis Bitter Magnet Laboratory, Massachusetts Institute of Technology, Cambridge, MA 02139
**Department of Electrical Engineering and Computer Science, Massachusetts Institute of Technology, Cambridge, MA 02139

ABSTRACT

Several approaches for achieving reversible hydrogen uptake by carbon are considered, including intercalation, adsorption by a graphite surface, hydrogenation of fullerenes, and the filling of carbon nanotubes. Most scenarios suggest that it is difficult to achieve an atomic uptake [H/C] ratio exceeding unity. Evidence for H_2 uptake by various carbon materials is reviewed.

INTRODUCTION

The objective of this paper is to consider the maximum amount of hydrogen that can be taken up reversibly by carbon-based materials. By reversibility, we mean that the hydrogen will be easily loaded into the carbon host material during the charging cycle, and readily unloaded during the discharge cycle. Several approaches are considered, including intercalation, adsorption on a single graphite surface or on multiple surfaces, hydrogen uptake by fullerenes, and by carbon nanotubes. Irreversible approaches are not considered, such as the uptake of hydrogen to form methane CH_4. Although the atomic [H/C] ratio is 4 for CH_4, hydrogen is strongly bonded to carbon in this compound and cannot easily be removed in a reversible process. Furthermore, catalytic and other chemical methods, that might be necessary for actually implementing the hydrogen uptake for any of these approaches, are not discussed. By considering these approaches from a broad perspective, we hope to identify the most attractive approach (or approaches) for hydrogen uptake by carbon, which should then be considered in more detailed studies.

GENERAL CONSIDERATIONS

In the ground state, the hydrogen molecule is very nearly spherical and the intermolecular interactions are weak [1]. The highest hydrogen concentration possible under atmospheric pressure is achieved by packing the H_2 molecules in a solid with a close-packed configuration, such as its low temperature solid phase. The structure of solid hydrogen at 4.2 K is hexagonal close-packed (space group $P6_3/mmc$), with lattice parameters $a = 3.76$ Å and $c = 6.14$ Å [2]. Thus the c/a ratio is 1.633, which is essentially the value of $c/a = \sqrt{8/3} = 1.633$ for the ideal hexagonal close-packing of spheres. The density of solid hydrogen is 0.071 g/cm³ which is about 3.5 times more dense than hydrogen gas in an advanced compressed gas cylinder [3]. In the ground rotational state, the H_2 molecule is approximately spherical in shape with a kinetic diameter of 2.97 Å [4].

Using purely geometric arguments, we can thus gain an estimate of the ideal stacking

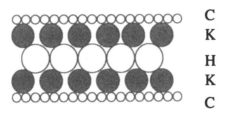

C
K
H
K
C

Figure 1: Schematic diagram of a ternary first stage C_4KH_x compound [9].

capacity of hydrogen molecules above a plane of graphite, which has a honeycomb structure (also $P6_3/mmc$), with lattice parameters $a = 2.46$ Å and $c = 6.71$ Å. Since the value of a for graphite is larger than the diameter of the hydrogen molecule, the closest packing of an adsorbed hydrogen layer would have to be incommensurate with the graphite planar surface. An atomic ratio of [H/C] $\sim$1 is considered to be the minimum uptake of practical interest for energy related applications [5]. Because of the 12:1 ratio between the masses of carbon and hydrogen, an atomic ratio of [H/C] $= 1$ corresponds to 8.3 wt% of hydrogen to carbon.

GRAPHITE INTERCALATION COMPOUNDS

The intercalation process involves the insertion of atoms or molecules of various guest species between planes of graphite which acts as a host material. The stage index (n) is used to describe the graphite intercalation compounds (GIC), where n is the number of planes of graphite separating one intercalate layer from another [6]. As a result of intercalation, the graphite sample expands along the c-axis to accommodate the planes of intercalant lying between two graphene layers. This increase in elastic energy is balanced by the attractive electrostatic interaction between the intercalate and graphite layers because of charge transfer that occurs during the intercalation process [6]. The greatest intercalant uptake occurs at stage $n = 1$. Intercalation of fullerene films [7] and carbon nanotubes [8] have also been reported, and hydrogen uptake in fullerenes and carbon nanotubes are discussed below.

Unfortunately, hydrogen does not seem to intercalate directly into a graphite host material. Hydrogen can, however, be co-intercalated with an alkali-metal in a chemical intercalation process to obtain stage $n = 1$ or stage $n = 2$ compounds, while a physisorption process can be used for hydrogen uptake on a second stage alkali-metal intercalated graphite host. Chemisorption of hydrogen by graphite is apparently accompanied by a back-donation of electrons from the alkali metal to the graphene and hydrogen layers in an alkali-metal GIC (see Fig. 1). There are two ways of introducing hydrogen into graphite in a chemisorption process: intercalation of the alkali-metal hydride and adsorption of hydrogen gas into an alkali-metal GIC. The resulting intercalation compound has the stoichiometry $C_{4n}KH_x$ (n is the stage index) [9, 10], where the value of x varies from 0.67 to 0.8, depending on the synthesis method.

Regarding chemisorption, many studies have been conducted on potassium hydride (KH) intercalated into GICs [9–12]. KH is an ionic insulator, with the hydrogen existing as an anion (H^-). The resulting graphite intercalation compound can be stage $n = 1$ ($C_4KH_{0.8}$) or $n = 2$ ($C_8KH_{0.8}$), depending on the experimental growth conditions (i.e., temperature or exposure time). $C_4KH_{0.8}$ yields a [H/C] ratio of 20 atom%, while $C_8KH_{0.8}$ yields 10 atom%.

The structural model (see Fig. 1) for the intercalation compound $C_{4n}KH_x$ consists of an ionic K^+–H^-–K^+ triple layer, sandwiched between graphene sheets, with the hydrogen acting like a metallic layer sheet between the two K^+ intercalate layers [10, 12]. Thus the uptake of hydrogen in the chemisorption process results in a large increase in the c-axis repeat distance I_c, from a value of $I_c = 5.42$ Å in the stage 1 C_8K to the stage 1 hydrogenated compound $C_4KH_{0.8}$ where $I_c = 8.53$ Å [13]. This large increase in I_c arises from the fact that the intercalate layer for the binary compound C_8K is a single layer while that for $C_4KH_{0.8}$ is a triple layer sandwich [14] The hydrogen uptake for a stage 2 compound is approximately half the value of that for a stage 1 compound, yielding a correspondingly lower [H/C] ratio [15]. However, the atomic ratio of [H/K] for the intercalate layer itself remains the same for the stage 1 and stage 2 compounds.

Gaseous hydrogen can be chemisorbed into the binary graphite intercalation compound C_8K to form the ternary GIC hydride $C_8KH_{2/3}$ [15]. This stoichiometry allows for even less hydrogen uptake than for $C_4KH_{0.8}$, with a [H/C] ratio of 8.3% for $C_8KH_{2/3}$, the compound mentioned explicitly in the literature [14, 16]. In the case of $C_4KH_{2/3}$ the I_c value is also 8.53 Å, and the interpretation of this lower H uptake is that there are more hydrogen vacancies in the intercalate layer in this case. Reflectivity measurements indicate that, for both the hydride and gas phase chemisorption synthesis methods, the hydrogen $1s$ electronic state lies below the Fermi level E_F, and therefore the hydrogen in the intercalant layer exists as H^- and acts as an acceptor [11]. The alkali-metal is believed to contribute electrons to both the graphene and hydrogen layers. The back-donation of electrons to the hydrogen provides the mechanism for dissociation of the molecule to form the charged atomic layer.

The physisorption of hydrogen involves weak bonding to the graphene layers through weak van der Waals forces. This physisorption process, however, results in a small expansion of the c-axis lattice parameter of $C_{24}K$ by only 0.29 Å after hydrogen gas adsorption at liquid nitrogen temperatures. It has also been observed that isotopes of hydrogen are more easily adsorbed and stabilized than hydrogen itself [17, 18]. Reflectance studies [19] of H_2 physisorbed into $C_{24}K$ showed very little back donation of charge. A stoichiometry of $C_{22.6}K(H_2)_{1.6}$ was found for this physisorbed compound [19], yielding a [H/C] ratio of 14 atom%. For both chemisorbed and physisorbed GICs, alkali metal species are also present, so that the net amount of hydrogen per unit volume or per gram is low relative to the other hydrogen uptake approaches discussed below.

ADSORPTION OF HYDROGEN ON GRAPHITE SURFACES

The structure of physisorbed hydrogen on a graphite surface has been measured using neutron scattering [20]. For sparse coverage, the H_2 molecules initially adsorb in a triangular structure that is commensurate with the graphite structure. The H_2 molecules therefore sit above every third carbon hexagon on the surface in the ($\sqrt{3} \times \sqrt{3}$) structure (see Fig. 2). This yields a lattice parameter of 4.26 Å for the H_2 triangular lattice. The ratio of the number of hydrogen atoms to carbon [H/C] for a commensurate ($\sqrt{3} \times \sqrt{3}$) H_2 monolayer would be 1:3 or 33 atom%. At full monolayer coverage, the H_2 molecules form a dense triangular structure, which is incommensurate with the graphite (see Fig. 2), yielding an uptake ratio [H/C] = $[2.46/3.76]^2$ = 43 atom%. However, once a second layer begins to form, a lattice parameter of 3.51 Å is observed [21]. This is smaller than the value of 3.76 Å measured for bulk hexagonal close-packing of hydrogen molecules, so that the uptake of a full double layer would yield [H/C] = 98 atom%. Thus, the adsorption of hydrogen as a

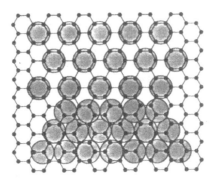

Figure 2: Relative density of $(\sqrt{3} \times \sqrt{3})$ commensurate (top) and incommensurate (bottom) monolayer of H_2 on a graphite surface.

densely packed double layer begins to sound interesting for energy-related applications.

Suppose that a special carbon fiber could be produced that would allow for stage $n = 1$ intercalation similar to the adsorption of hydrogen at a graphite surface. If lattice stability could be maintained when hydrogen uptake occurs to the solid hydrogen density at the upper and lower sides of each graphene sheet, we should be able to achieve quite a high hydrogen uptake of $[H/C] = 2 [2.46/3.51]^2 (6.71/6.14) = 107$ atom%. It would be a real challenge to the catalytic community to develop a method for synthesizing such a special carbon fiber with these capabilities, representing about the best scenario one can imagine for hydrogen uptake. If such high uptake were to occur, we would expect an expansion of the c-axis spacing (as could be studied by a diffraction experiment) by about a factor of about 2–3 to accommodate the double hydrogen layers between sequential carbon layers. Such a structure might well be stabilized by some amount of charge transfer between the hydrogen and carbon layers. If charge transfer occurred it should be observable, as a modification to either the C–C or H–H vibrational frequencies in a Raman scattering experiment.

HYDROGENATION OF FULLERENES

From elementary arguments, each C_{60} molecule contains 30 double bonds at the union between two hexagons (see Fig. 3). If we could imagine that at each double bond site an H_2 molecule is attached, this could give a possible model for the observation of hydrogenation of C_{60} to the reported stoichiometry of $C_{60}H_{60}$ [22]. The uptake of hydrogen at the C_{60} surface is related to both surface adsorption and intercalation, which are both discussed above. For the case of $C_{60}H_{60}$, the uptake ratio is high $[H/C] = 100$ atom%, and even for the compound $C_{60}H_{36}$, which has also been reported [23], we obtain $[H/C] = 60$ atom%. From these arguments we conclude that hydrogenation of the surface of fullerene cage molecules is quite promising as an approach for hydrogen-related energy applications, provided that these high uptake values can be achieved in the solid state and a method can be found to make the reaction reversible. It is not known that fully hydrogenated fullerenes $C_{60}H_{60}$ or even $C_{60}H_{36}$ can form a stable crystal. The lower density of the fcc C_{60} film (1.72 g/cm^3) compared to graphite (2.26 g/cm^3) would also have to be considered in comparing a hypothetical hydrogenated fullerene film to the other approaches considered in this review.

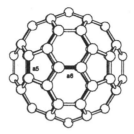

Figure 3: Model for C_{60} showing single (a_5) and double bonds (a_6) [7].

Figure 4: Model for an array of carbon nanotubes on a triangular lattice with a lattice constant of 17.1 Å [25].

FILLING OF CARBON NANOTUBES

The possibility of filling the hollow core of a carbon nanotube (single and multi-walled) with different guest species, and modifying its physical properties, has generated much interest from both a scientific and experimental standpoint [24]. In particular, carbon nanotubes (see Fig. 4) have been investigated as a possible storage mechanism for hydrogen with applications to fuel cell technology [24]. If we were to take a single rectangular sheet of graphite with lattice parameter $a = 2.46$ Å, and roll it into a seamless cylinder, we could perhaps generate a gas container of nanometer size, as suggested in Ref. [24]. If we consider a (10,10) armchair nanotube, then on a purely geometric basis, we could get nearly as much hydrogen adsorption for a monolayer on the inside and outside of this carbon nanotube as we would get on both sides of a single graphene sheet. Taking the tube curvature and the finite size of the hydrogen molecules into account, an estimate of about 33 atom% is obtained for [H/C] for monolayer coverage at the density of solid hydrogen on the inside of a (10,10) carbon nanotube. If hydrogen at this density were to fill the entire hollow of the (10,10) tube, a value for [H/C] of 73 atom% is estimated. If now hydrogen also fills the interstitial volume between the nanotubes (see Fig. 4), we then obtain an estimate of 41 atom% in the interstitial volume, so that if hydrogen could be taken up both inside and outside of (10,10) nanotubes in a triangular lattice with a lattice constant of 17.1 Å and an inter-tube separation of 3.1 Å, then the total uptake would be [H/C] = 114 atom%. By varying the growth conditions of single-wall nanotube ropes their mean diameters can be varied. Diameter values up to 2 nm have been reported [26]. For these larger 2 nm diameter nanotubes, the amount of hydrogen that could be stored within the nanotube core at the solid hydrogen density would be increased to [H/C] $\sim$250 atom%. This is a very optimistic value for [H/C] based on an idealization of present technology.

Dillon *et al.* [24] investigated the hydrogen adsorption capacity of an experimental sample of single-walled carbon nanotubes (which also contained some amorphous carbon) in comparison with that of activated carbon. Through temperature programmed desorption (TDP) measurements, they found that hydrogen desorbed from the two sets of as-prepared samples at about the same temperature ($\sim$133 K). If they heated the nanotube sample in vacuum, and thus opened the ends of the tubes, a second peak appeared at a higher temperature ($\sim$290 K) in the TDP vs. temperature spectrum, indicating that there was now a second type of adsorption site. They attributed the new peak to desorption of H_2 that had physisorbed inside the hollow core of the carbon nanotubes. They note that this high temperature desorption site has not been seen in either activated carbon samples or in arc-generated soots produced without a catalyst [24]. The hydrogen uptake capacity of this high temperature site was found to be 5–10 wt%, or an [H/C] ratio of about 60–120 atom%.

CONCLUSIONS

Several approaches have been identified which could under optimistic scenarios yield an atomic [H/C] ratio of about 100 atom%, including double layered hydrogen interstitially arranged between graphene layers in specially designed carbon fibers, hydrogenated fullerenes and hydrogen-loaded carbon nano-cylinders, considering the hydrogen density in all cases to be that of solid hydrogen. Therefore it appears unlikely that hydrogen densities greater than solid molecular hydrogen could be achieved under ambient conditions, unless unforeseen developments lead to the uptake of additional hydrogen layers (beyond 2) on an internal graphite surface. Some increase in the [H/C] ratio could perhaps be achieved by increasing the diameter of the single-wall nanotubes, or by finding a method for multi-layer loading of these special carbon fibers, beyond the double layers considered here.

REFERENCES

[1]   J. K. Kranendonk. In *Solid Hydrogen*. Plenum Press, New York, 1983.

[2]   In *Landolt-Börnstein Numerical Data and Functional Relationships in Science and Technology, New Series III/14a*, edited by K.-H. Hellwege and A. M. Hellwege, page 18. Springer-Verlag, Berlin, 1988.

[3]   S. Hynek, W. Fuller, J. Bentley, and J. McCullough, in *Proc. of the 10th World High Energy Conference*, **2**, 985–1000 (1994).

[4]   J. Koresh and A. Soffer, J.C.S. Faraday Trans. I. **76**, 2472–2485 (1980).

[5]   M. DeLuchi, in *Hydrogen fuel-cell vehicles*, (Institute of Transportation Studies, Univ. of California, Davis, 1992).

[6]   M. S. Dresselhaus and G. Dresselhaus, Adv. Phys. **30**, 139 (1981).

[7]   M. S. Dresselhaus, G. Dresselhaus, and P. C. Eklund, *Science of Fullerenes and Carbon Nanotubes* (Academic Press, New York, NY, 1996).

[8] A. M. Rao, P. C. Eklund, S. Bandow, A. Thess, and R. E. Smalley, Nature (London) **388**, 257 (1997).

[9] D. Guerard, C. Takoudjou, and F. Rousseaux, Synth. Met. **7**, 43 (1983).

[10] T. Enoki, K. Nakazawa, K. Suzuki, S. Miyajima, T. Chiba, Y. Iye, H. Yamamoto, and H. Inokuchi, J. Less-Common Metals **172-174**, 20 (1991).

[11] G. L. Doll, M. H. Yang, and P. C. Eklund, Phys. Rev. B **35**, 9790 (1987).

[12] S. Miyajima, M. Kabasawa, T. Chiba, T. Enoki, Y. Maruyama, and H. Inokuchi, Phys. Rev. Lett. **64**, 319 (1990).

[13] D. Guérard, G. M. T. Foley, M. Zanini, and J.E. Fischer, Nuovo Cim. **38**, 410 (1977).

[14] M. S. Dresselhaus and G. Dresselhaus, Advances in Phys. **30**, 139–326 (1981).

[15] G. L. Doll and P. C. Eklund, J. Mater. Res. **2**, 638 (1987).

[16] P. Lagrange and A. Hérold, Carbon **16**, 235–240 (1978).

[17] T. Terai and Y. Takohaski, Synth. Met. **7**, 49 (1983).

[18] G. L. Doll, P. C. Eklund, and G. Senatore. In *Intercalation in Layered Materials*, edited by M. S. Dresselhaus. Plenum Press, New York, 1986.

[19] G. L. Doll and P. C. Eklund, Phys. Rev. B **36**, 9191 (1987).

[20] M. Nielsen. In *Phase Transitions in Surface Films*, edited by J. G. Dash and J. Ruvalds. Plenum Press, New York, 1980.

[21] M. Nielsen, J. P. McTague, and W. Ellenson, J. Phys. **38**, C4/10 (1977).

[22] D. Koruga, S. Hameroff, J. Withers, R. Loutfy, and M. Sundareshan, in *Fullerene C₆₀: History, Physics, Nanobiology, Nanotechnology*, (North-Holland, Amsterdam, 1993).

[23] R.E. Haufler, J. J. Conceicao, L. P. F. Chibante, Y. Chai, N. E. Byrne, S. Flanagan, M. M. Haley, S. C. O'Brien, C. Pan, Z. Xiao, W. E. Billups, M. A. Ciufolini, R. H. Hauge, J. L. Margrave, L. J. Wilson, R. F. Curl, and R. E. Smalley, J. Phys. Chem. **94**, 8634 (1990).

[24] A. C. Dillon, K. M. Jones, T. A. Bekkedahl, C. H. Kiang, D. S. Bethune, and M. J. Heben, Nature **386**, 377 (1997).

[25] A. Thess, R. Lee, P. Nikolaev, H. Dai, P. Petit, J. Robert, C. Xu, Y. H. Lee, S. G. Kim, A. G. Rinzler, D. T. Colbert, G. E. Scuseria, D. Tománek, J. E. Fischer, and R. E. Smalley, Science **273**, 483–487 (1996).

[26] J. W. G. Wildöer, L. C. Venema, A. G. Rinzler, R. E. Smalley, and C. Dekker, Nature (London) **391**, 59–62 (1998).

Part IV

Pillared Layered and Porous Catalysts

MACROPOROUS MATERIALS WITH UNIFORM PORES BY EMULSION TEMPLATING

A. IMHOF, D.J. PINE
Department of Chemical Engineering and Department of Materials, University of California, Santa Barbara, California 93106-5080.

ABSTRACT

A method was developed for the production of macroporous oxide materials by using the droplets of a nonaqueous emulsion as the templates around which material is deposited through a sol-gel process. Moreover, uniform pores arranged in regular arrays can be obtained by starting with an emulsion of uniform droplets. These droplets first self-assemble into a colloidal crystal after which gelation of the suspending sol-gel mixture captures the ordered structure. After drying and calcination pellets are obtained which contain ordered arrays of spherical pores left behind by the emulsion droplets. The method can be used to make uniform pores in the range from 0.05-5 micrometers in many different materials. We demonstrate the process for titania, silica, and zirconia.

INTRODUCTION

There is intense interest in producing porous materials with highly uniform pore sizes for use as catalytic materials, filters, and adsorbents. Methods exist for the fabrication of materials with uniform pores of diameter less than ca. 20 nm [1,2]. They are formed by templating surfactant molecules with inorganic solutes such as silicates. However, there is no general method for producing such materials with uniform pore sizes at larger length scales. Existing processes for producing macroporous materials result in a very broad distribution of pore sizes [3,4]. Here, we report a new method for producing macroporous materials with highly uniform pores in the size range from 50 nm to several micrometers. A consequence of the high uniformity of the pores is that they order into crystalline lattices. This could make these materials also useful in optical devices such as photonic band gap materials [5,6].

The basic strategy is to use the droplets in an emulsion as the inert templates for a sol-gel process that takes place in the continuous (external) phase of the emulsion. Emulsions of equally-sized droplets can be produced through a repeated fractionation procedure [7]. Such monodisperse droplets undergo a spontaneous transition to an ordered colloidal crystalline phase [8]. The sol-gel process is then used to produce a space-filling gel of one of a range of possible solid materials. This gel encapsulates the droplets, thus forming 'imprints' of the them in the resulting material. The major advantage of templating emulsion droplets (as opposed to solid microparticles) is that the droplets can easily be removed by evaporation or dissolution *prior to* drying of the gel. Since drying is accompanied by significant shrinkage we found that this feature is essential in order to obtain uncracked materials. A further advantage is that the deformability of droplets allows emulsion volume fractions in excess of the close packing limit for hard spheres (74%). This enables the fabrication of porous materials with filling fractions lower than 26%. Such materials have a connected pore system.

One of the problems that has to be overcome to make this strategy work is that sol-gel processes in general make use of metal alkoxides dissolved in an alcohol and hydrolyzed by the controlled addition of water. Conventional emulsions, however, are aqueous systems and are thus incompatible with most metal alkoxides. We therefore first developed a suitable nonaqueous emulsion [9] in which the water is replaced by another polar liquid, namely formamide. This led to stable oil-in-formamide emulsions. Then, the sol-gel process was adapted so that it would work in the highly polar formamide as the solvent instead of the usual lower alcohols.

We demonstrate the emulsion templating process by making ordered macroporous titania, silica, and zirconia. It should, however, not be difficult to adapt the process to making a range of other metal oxides, binary metal oxides, and even organic polymer gels.

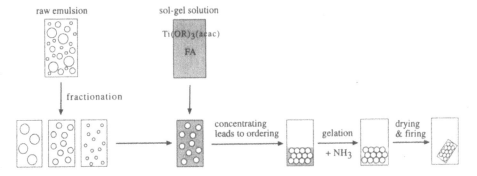

Figure I. Schematic of the emulsion templating process.

EXPERIMENT

The emulsion templating process is shown schematically in Figure I and is further detailed below.

Emulsion preparation

In ref. [9] it was found that stable nonaqueous emulsions can be prepared with formamide (FA) as the continuous phase. A surfactant that gives optimal stability was the symmetric triblock copolymer (ethylene glycol)$_{20}$-(propylene glycol)$_{70}$-(ethylene glycol)$_{20}$. Its molecular weight is 5800 and it contains 30% by weight ethylene glycol monomer. For the emulsion templating we used oil that consisted of 99% iso-octane and 1% silicone oil. The silicone oil (or another high molecular weight oil) is needed to prevent droplet coarsening due to Ostwald ripening [9].

In a typical recipe 20 g of surfactant was dissolved in 1 liter of formamide. Then 150 ml of oil was emulsified in it using a homogenizer. The emulsion was fractionated with the method described in ref. [7]. This resulted in a dozen or so monodisperse emulsions of between 1 and 3 ml in size, each having an oil volume fraction of about 50%.

Sol-gel precursor solution

A typical preparation is as follows: 26 mmol of titanium tetraisopropoxide (97%) was treated with an equimolar amount of the chelating agent acetylacetone (acac) to reduce its reactivity with water. The resulting complex is insoluble in formamide but it can be made soluble by partial hydrolysis at a H_2O/Ti ratio of between 3 and 10. Thus, the solution was mixed by vigorous stirring with a mixture of 1.6 g of water in 6.6 g of formamide. The resulting clear yellow solution contains a considerable amount of isopropanol produced by the hydrolysis reaction. Since isopropanol will destabilize the emulsion, it was removed by extracting twice with a fivefold excess of hexanes. The resulting yellow liquid was frequently turbid. Briefly heating it to ~90°C produced a clear yellow and slightly viscous solution. Its color changed slowly from yellow to orange over the coarse of a few days but it did not form a precipitate for several weeks. The solution was used for templating within a few days. We found that it forms a clear rigid gel a few hours after mixing it with a small amount of 30% ammonia.

A sol-gel solution for zirconia was made in the same way, starting with a 70% zirconium tetra-n-butoxide solution in n-butanol. A silica precursor solution was made by mixing 4 ml of silicon tetramethoxide into a mixture of 1 ml 0.1 M HCl and 5 ml formamide, and the hexane extraction was not used.

Emulsion templating

The emulsion droplets were then transferred to the sol-gel solution. First, 2 wt% of the surfactant was dissolved in the sol-gel solution for droplet stabilization. Then, 1 ml of a 50% v/v emulsion of monodisperse droplets was mixed with 3 ml of the sol-gel solution. The droplets were centrifuged at 1500 rpm and collected as a cream. This was then redispersed in another 3 ml of the sol-gel solution, and centrifuged again. The resulting concentrated emulsion was transferred to a polyethylene vial and mixed with a tiny amount of 30% ammonia from a syringe such that the molar ratio $NH_3/Ti \approx 1$. This induced gelation in about 3 hours, which is enough time for the droplets to self-assemble into a crystalline lattice whenever the droplets take up more than ~50% of the final emulsion volume [8]. About 16 hours after gelation the gel was aged at 50°C for 24 hours. Then it was immersed in plenty of ethanol for 24 hours in order to extract formamide, iso-octane, and surfactant, all of which are all soluble in ethanol. The gel was then dried slowly at room temperature over several days and finally calcined in air in a furnace at temperatures of up to 1000°C for 2 hours with a heating rate of 15°C/min.

Analyses

The porous materials were examined with a JEOL 6300F Scanning Electron Microscope (SEM). Samples were prepared by breaking a small chip off the pellets and polishing the exposed, internal surface with diamond paper and sputtering it with gold. X-ray diffractograms were taken with a Philips X'pert-MPD using Cu-Kα radiation. Nitrogen sorption isotherms were measured at 77 K with a Micromeritics ASAP2000 on samples degassed at 200°C for 4 hours. Thermogravimetric analysis (TGA) was done with a Mettler TA3000 on gels dried at 60°C.

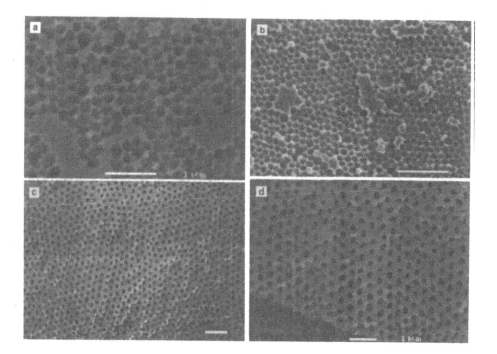

Figure II. SEM images of emulsion templated titania showing ordered macropores. *a*. After drying at 60°C; *b*. Same sample after calcination at 1000°C; *c*. Highly ordered sample calcined at 1000°C; *d*. Same as *c* showing an edge with cusps. Scale bars are 1 μm.

RESULTS

The emulsion templating procedure yielded white pellets that were usually uncracked. The whiteness is due to strong scattering of light by the macropores which are on the order of the wavelength of visible light.

A SEM picture of a dried (but not calcined) macroporous titania gel is shown in Figure IIa. The droplets have left behind uniform spherical pores of 175 nm that show ordering into hexagonal domains. The original emulsion had a droplet volume fraction of about 55%. During the drying process the gel shrunk to about 50% of its wet size. Thus, the templating process was successful. The pore structure did not collapse, even though drying generally produces very large stresses in the gel. This is due in large part to the fact that the template was removed *before* drying. It is possible that residual surfactant in the gels also helped to reduce the stresses by lowering the surface tension of the menisci that form during drying.

In order to remove residual organic material and to permit densification of the titania matrix most gels were given a heat treatment. In Figure IIb the same sample is shown as in Figure IIa after being calcined at 1000°C. The titania has become a little more 'grainy', but the ordered pore structure has remained completely intact. Due to the densification of the titania matrix has shrunk the pores to 145 nm but apparently these processes did not damage the pore structure.

Figure IIc shows a titania sample with a high degree of pore order. The pores appear not to be touching, but this is caused by the fact that the pores were 'decapitated' by the surface. Their actual diameter is closer to the average distance between pore centers (380 nm). This can be seen in the bottom left hand corner of Figure IId, where the edge cuts through the middle of the pores showing cusps corresponding to the size of the macropores. These images also show that the pore lattice exhibits normal lattice defects such as vacancies and dislocations.

A typical TGA trace of a titania gel is seen in Fig. III. These samples typically lost 30-35% of their weight during the heat treatment. Above 500°C weight loss is essentially complete. Most of the loss is probably due to the oxidation of unhydrolysed acetylacetonato groups and to evaporation of formamide which has a high boiling point of 210°C.

The densification of the TiO_2 matrix was further studied by measuring nitrogen sorption isotherms. The isotherms up to 800°C were of type IV and showed mesoporosity in the range 2-10 nm. At 900°C the isotherms changed to type II, which indicates that no pores smaller than ca. 50 nm exist. From the data the BET specific surface area (S_{BET}) of the titania samples was calculated as well as the total volume of pores smaller than 50 nm. Note that this 'mesoporosity'

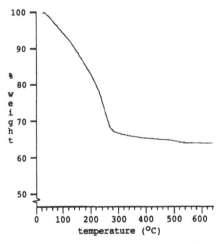

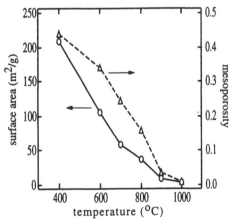

Figure III. Thermogravimetric analysis of a macroporous titania gel.

Figure IV. Specific surface area (circles) and mesoporosity (triangles) of macroporous titania.

170

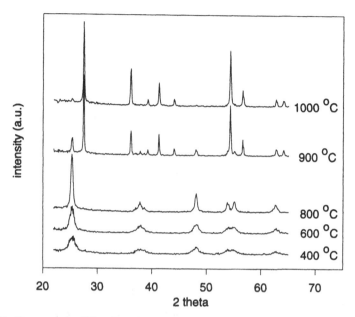

Figure V. X-ray powder diffraction of titania matrix after 2 hour heat treatment at the temperatures shown. The anatase structure changes to rutile at 900°C.

therefore does not include the volume of the macropores formed by the templating mechanism. The results are shown in Figure IV and show a strong densification of the TiO_2 matrix in the range 400-1000°C. This kind of control is useful because it can be done without harming the macropore structure. In applications such as in catalytic materials, for example, the macropores facilitate transport to mesoporous internal regions where reactions can take place.

The crystallinity of the TiO_2 matrix as a function of calcination temperature was investigated by X-ray diffraction; see Figure V. Diffraction maxima first appear at around 400°C and are quite broad, indicating very small TiO_2 crystallites. We calculated an average crystallite size of only 7 nm using the Scherrer formula. At higher temperatures the peaks sharpen due to grain growth. The diffraction pattern corresponds to TiO_2 in the anatase structure. At 900°C the titania recrystallized to rutile, but still contained some anatase. Above that temperature only rutile was present and the grains had grown to 60 nm. It was satisfying to see in SEM images, however, that all these microstructural changes affected only the matrix material but had no effect upon the ordered pore structure. Only at 1100°C did the pore structure begin to degrade as a result of excessive grain growth.

In order to demonstrate the versatility of the emulsion templating technique we also made ordered macroporous zirconia and silica. SEM images of these samples are shown in Figure VI. The silica sample had a high porosity of 89%. This was only possible due to deformation of the emulsion droplets in the original emulsion. As a result the voids in the gel overlap, thus creating a continuous pore structure.

CONCLUSIONS

Templating of nonaqueous emulsions is a new and versatile technique that allows one to fabricate macroporous materials with ordered pores. It is also inexpensive, yields uncracked materials, and

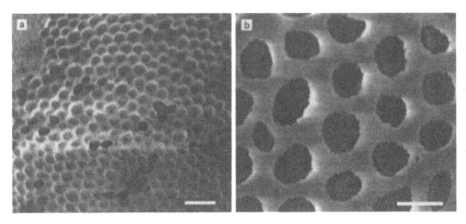

Figure VI. *a.* SEM image of emulsion templated ZrO_2 after calcination at 1000°C. *b.* SEM image of emulsion templated SiO, calcined at 600°C. Scale bars are 1 μm.

can be applied to a wide variety of solids. Of course, in applications where a uniform pore size is unimportant emulsion templating still has these advantages, but other methods exist. When size uniformity is desired, however, emulsion templating is the only method available to produce macroporous materials. The most severe limitation currently is the production of large quantities of monodisperse emulsion. The fractionation process that we have used thus far is time consuming and yields only small amounts of material. Thus, it would be desirable to develop a more efficient means for the production of large quantities of monodisperse emulsions. Developing schemes for producing highly ordered droplets in the emulsion prior to gelation is also a high priority. Promising routes include application of shear flow and external fields, and droplet deposition on surface microstructures [10,11].

REFERENCES

1. C.T. Kresge, M.E. Leonowicz, W.J. Roth, J.C. Vartuli, and J.S. Beck, J. Am. Ceram. Soc. **114**, 10834 (1992).
2. Q. Huo, D.I. Margolese, U. Ciesla, P. Feng, T.E. Gier, P. Sieger, R. Leon, P.M. Petroff, F. Schüth, and G.D. Stucky, Nature **368**, 317 (1994).
3. M. Wu, T. Fujiu, and G.L. Messing, J. Non-Cryst. Solids **121**, 407 (1990).
4. R.L. Downs, M.A. Ebner, and W.J. Milner in Sol-Gel Technology for Thin Films, Fibers, Preforms, Electronics, and Specialty Shapes, edited by L.C. Klein (Noyes Publications, Park Ridge, NJ, 1988, pp.330-381.
5. E. Yablonovitch, J. Opt. Soc. Am. B **10**, 283 (1993).
6. Photonic Band Gap Materials, edited by C. Soukoulis (Kluwer, Dordrecht, 1996).
7. J. Bibette, J. Colloid Interface Sci. **147**, 474 (1991).
8. See e.g. P.N. Pusey and W. van Megen, Nature **320**, 340 (1986).
9. A. Imhof and D.J. Pine, J. Colloid Interface Sci. **192**, 368 (1997).
10. A.D. Dinsmore, A.G. Yodh, and D.J. Pine, Nature **383**, 239 (1996).
11. A. van Blaaderen, R. Ruel, and P. Wiltzius, Nature **385**, 321 (1997).

THE HYDROTHERMAL SYNTHESIS AND CHARACTERIZATION OF NEW ORGANICALLY TEMPLATED LAYERED VANADIUM OXIDES BY METHYLAMINE

Rongji Chen, Peter Y. Zavalij, and M. Stanley Whittingham*
Chemistry Department and Materials Research Center
State University of New York at Binghamton, Binghamton, New York 13902-6016

ABSTRACT

The hydrothermal reaction of vanadium pentoxide with methylamine leads to a series of new layered vanadium oxides by using different acids to adjust the initial pH. The structure of these layered vanadium oxides was established by X-ray single crystal and powder diffraction. The new phases were also characterized by TGA and FTIR. The oxidation state of vanadium was determined. These materials are expected to exhibit catalytic activity and may be of interest as cathode materials for secondary lithium batteries.

INTRODUCTION

The synthesis and characterization of new materials by soft chemistry approaches has been of much interest recently [1]. The materials formed under kinetic conditions are usually metastable with different structures and properties. Hydrothermal reactions, which happen naturally in the formation of minerals in the crust of the earth, were first utilized to synthesize zeolites and phosphates, both as catalysts. Hydrothermal synthesis under mild conditions has been shown as one of the most successful methods to synthesize these materials. Using organic templates with hydrothermal reaction has been shown to be very successful in the formation of new metal oxides with open tunnel structures [2], layered vanadium oxides with organic templates in-between the layers [3-5] and more recently some mixed transition metal oxides [6]. Although the tetramethylammonium (TMA) ion is one of the most commonly used in hydrothermal synthesis, other amines [7] or surfactants [8] have also been found to be very adept in forming new structural compounds.

In this paper, we report the formation of new layered vanadium oxides phases containing organic amines, in which the use of the organic template methylamine was the critical element. Previously we reported for both tungsten oxides [9] and vanadium oxides [3] the criticality of controlling the pH of the reaction medium, as it determines the phase formed. The initial pH of the reaction media can be adjusted by strong mineral acids such as HCl/HNO_3 or by a weak acid such as acetic acid, HAc. The advantage of acetic acid is that, in the presence of the TMA ion, it behaves as a buffer. The pH of the reaction media remains almost constant during the whole hydrothermal heating. These new compounds consist of vanadium oxide layers with methylamine residing between the layers. These materials are expected to exhibit catalytic activity, and might well also be active in electrochemical processes.

EXPERIMENTAL

Three layered vanadium oxides were hydrothermally synthesized by dissolving V_2O_5 powder from Johnson & Matthey in methylamine aqueous solution from Fisher Scientific in a 1:2 molar ratio. After adjusting the pH with an acid, the final solution was transferred to a 125-ml Teflon lined autoclave (Parr bomb), sealed, and reacted hydrothermally for 4 or 5 days at 200°C. When HAc was used to adjust the pH to 4, a block of dark green clay $(CH_3NH_3)_{0.75}V_4O_{10} \cdot 0.67H_2O$ was obtained; the final pH after reaction was 4.5. When HAc was used to adjust initial pH to 7, large dark brown plate-like crystals $(CH_3NH_3)V_3O_7$ were obtained; the final pH was 7 after reaction. When HNO_3 was used to adjust the pH value to 3.5, black shining crystals $(CH_3NH_2)_2V_8O_{17}$ were obtained; the final pH after reaction was 9. All the compounds were overnight dried in 45°C oven before use.

X-ray powder diffraction data was recorded using $CuK\alpha$ radiation on a Scintag XDS 2000 $\theta-\theta$ diffractometer. The data was collected in a continuous scan at 0.1°/min over the range $6° < 2\theta < 70°$. The single crystal diffraction was performed on a Siemens Smart CCD diffractometer at Syracuse University. All crystallographic refinement was conducted using CSD (Crystal Structure Determination) package [10]. Thermal gravimetry (TG) data was obtained on a Perkin Elmer model TGA 7 at a heating rate of 3°C/min under oxygen. The FTIR spectrum was obtained by measuring a KBr pellet containing about 1% sample on Perkin Elmer 1600 series.

The oxidation state of the vanadium was determined by oxidizing all the vanadium to V^{5+} with $Ce(SO_4)_2 \cdot H_2SO_4$ following by the titration of the excess $Ce(SO_4)_2 \cdot H_2SO_4$ with $(NH_4)_2Fe(SO_4)_2$ using Ferrion as indicator. The N% was determined by the Kjeldahl method.

RESULTS AND DISCUSSION

X-Ray Diffraction Determination

The powder X-ray diffraction patterns of the three vanadium phases are shown in Figure 1. $(CH_3NH_3)V_3O_7$ was indexed in monoclinic system with space group $P2_1/c$ and the lattice parameters a=11.834(8) Å, b=6.663(4) Å, c=15.193(9) Å and ß=138.104(1)°. $(CH_3NH_3)_{0.75}V_4O_{10} \cdot 0.67H_2O$ was indexed in monoclinic system with space group C2/m and the lattice parameters a=11.6728(3) Å, b=3.6681(1) Å, c=11.0949(3) Å and ß=99.865(5)°. We are still working on the structure of $(CH_3NH_2)_2V_8O_{17}$.

The crystal structure of $(CH_3NH_3)V_3O_7$ (Fig. 2a) is built by V_3O_7 layers (Fig. 3a). Methylammonium cations are incorporated between V_3O_7 layers making 4 hydrogen bonds N-H---O with length from 2.88 to 3.03 Å whereas length of 4 contacts C-H---O are in range from 3.24 to 3.33 Å. Coordination polyhedra of vanadium atoms are square pyramids and tetrahedra. There are two types of square pyramids (SP) – corner SP that shares two edges next to each other and middle SP that shares opposite edges. The double bonded oxygen at the SP apex is not shared. The square pyramids (SP) sharing edges form chains. The corner and middle SP make a pair with the same apex orientation, and the next pair has the opposite

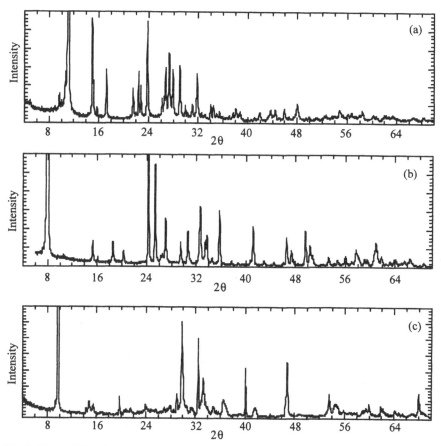

Fig. 1. X-ray diffraction patterns of the vanadium oxide phases, using CuKα radiation; (a) $(CH_3NH_3)V_3O_7$, (b) $(CH_3NH_3)_{0.75}V_4O_{10} \cdot 0.67H_2O$ and (c) $(CH_3NH_2)_2V_8O_{17}$.

orientation. The chains are joined in layers by corner shared tetrahedra (T). Each tetrahedron shares 2 corners with one chain and 1 with another, one corner remains free. The crystal structure of $(CH_3NH_3)_{0.75}V_4O_{10} \cdot 0.67H_2O$ is also layered but built by double V_2O_5 layer (fig. 2b). This type of layer was found in compounds intercalated with metals cations such as Ag^+ [11], K^+ [12, 13], Na^+ [14], as well as in xerogel V_2O_5 [15]. The methylammonium cation and water molecules are statistically disordered between the vanadium oxide layers. Coordination polyhedra of vanadium atom is distorted octahedron with double bonded oxygen in one apex and weekly bonded oxygen in opposite site.

The V_3O_7 layer found in $(CH_3NH_3)V_3O_7$, fig. 3(a), is similar to those found in other vanadium oxide compounds $(dabco)V_6O_{14} \cdot H_2O$ [16, 17], $(en_2M)V_6O_{14}$ (M=Zn, Cu) [18] and

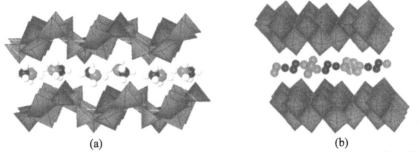

(a)   (b)

Fig. 2. The structure of (a) $(CH_3NH_3)V_3O_7$, and (b) $(CH_3NH_3)_{0.75}V_4O_{10} \cdot 0.67H_2O$, from [19].

(tma)V_3O_7 [20] (fig. 3b, 3c, 3d respectively) but detailed crystallographic analysis shows [19] that all four of them differ in the relative orientation of the SP chains and in the way the tetrahedra share their corners. The conformation of SP chain with tetrahedra attached by two corners (SP$_2$T) is the same in all compounds. The bases of both middle and corner SP as well as one face of tetrahedron are approximately parallel to each other. One of the other tetrahedron corners is shared with the adjoining SP$_2$T chain forming V_3O_7 layers. When the oxygen atom which lies in the same plane as the SP bases is shared, the V_3O_7 layer is flat (Fig. 3c, 3d). Otherwise (Fig. 3a, 3b) the layer is tilted and the distance between chains is significantly shorter than that in previous case. The second distinguishing feature between the layers is the relative orientation of the neighboring SP$_2$T chains. In the case of (dabco)V_6O_{14} and (tma)V_3O_7 (Fig. 3b, 3d) they are transformed each into another by simple translation in the direction perpendicular to the chain so that the repeat distance in that direction is equal to the distance between the chains. The transformation between chains can be more complex than simple translation. For instance in (ma)V_3O_7 and (en$_2$Zn)V_6O_{14} (Fig. 3a, 3c) two chains interact by glide plane perpendicular to the chain – translation and reflection. All these cases are summarized in the table below, where t is the distance between the chains [19]:

Table 1. Relationship between the four V_3O_7 structures.

Symmetry relations between SP$_2$T chains	Corner of tetrahedron shared between chains	
	Top ($\rightarrow$ tilted layer)	Bottom ($\rightarrow$ flat layer)
Glide plane	$(CH_3NH_3)V_3O_7$ (Fig. 3a) t=7.597 Å, c=2t	(en$_2$M)V_6O_{14} (Fig. 3c) t=7.851 Å, c=2t
Translation	dabcoV_6O_{14}·H$_2$O (Fig. 3b) t=7.574 Å, c=t	tmaV_3O_7 (Fig. 3d) t=8.427 Å, c=t

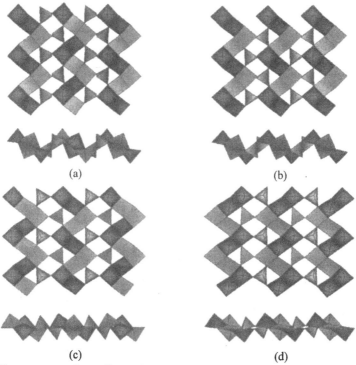

(a)   (b)

(c)   (d)

Fig. 3. The structures of vanadium oxide layer of (a) $(CH_3NH_3)V_3O_7$, (b) dabcoV_6O_{14}•H_2O, (c) $(en_2M)V_6O_{14}$, and (d) tmaV_3O_7, after [19].

Thermal Gravimetric Analysis

Thermogravimetric analyses of these three compounds under oxygen at a heating rate of 3°C/min are shown in Figure 4. $(CH_3NH_3)V_3O_7$ (Figure 4a) loses 1% surface water first. From 160°C, this compound burns up CH_3NH_2 and takes half of oxygen at the same time vanadium is oxidized and converts to V_2O_5 at 300°C. The total weight loss is 8.5% which matches very well with the theoretical calculation. Thermal decomposition of $(CH_3NH_3)_{0.75}V_4O_{10}$•$0.67H_2O$ is shown in Figure 4b. The TGA curve of this compound under O_2 indicates 3% surface or crystalline water loss following a 6% CH_3NH_2 loss before it converts to V_2O_5. $(CH_3NH_2)_2V_8O_{17}$ under oxygen (Figure 4c) experiences a total weight loss of 1.8% and finally decomposes to V_2O_5 at 550°C after several stages of weight loss and gain. There is no apparent boundary between each step. All the residues after thermal analysis are orange and their x-ray diffraction patterns show the presence of pure V_2O_5.

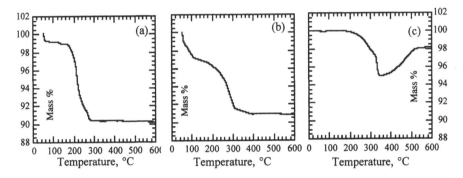

Fig. 4. Thermogravimetric analysis of the vanadium oxide phases at a heating rate of 3°C/min under O_2, (a) $(CH_3NH_3)V_3O_7$, (b) $(CH_3NH_3)_{0.75}V_4O_{10} \cdot 0.67H_2O$, and (c) $(CH_3NH_2)_2V_8O_{17}$.

FTIR Spectra

The FTIR spectrum of vanadium pentoxide (Figure 5d) shows three peaks at 1021 cm^{-1}, 827 cm^{-1} and 617 cm^{-1}, which have been associated with the V=O and two V-O vibrations, one bridging (900-700 cm^{-1}) and the other (800-400 cm^{-1}) to oxygens in chain positions between three vanadiums. Each of these three compounds (Figure 5a-c) not only have vanadium and oxygen bond vibration peaks (below 1100 cm^{-1}), but also has a peak at 1382 cm^{-1}, which belongs to the C-N vibration from $CH_3NH_3^+$ or CH_3NH_2. Both VO_5 square pyramids and VO_4 tetrahedra are present in the structure of $(CH_3NH_3)V_3O_7$, resulting in a greater number of different vanadium - oxygen bonds giving the richer IR spectrum, compared with $(CH_3NH_3)_{0.75}V_4O_{10} \cdot 0.67H_2O$ which contains just distorted VO_6 octahedra. The much broader C-N vibration observed in $(CH_3NH_2)_2V_8O_{17}$, suggests that the nitrogen lone pair electrons might be bonding to the vanadium, Figure 5c. In contrast, $CH_3NH_3^+$ in $(CH_3NH_3)V_3O_7$ and $(CH_3NH_3)_{0.75}V_4O_{10} \cdot 0.67H_2O$ has a lot of freedom, thus the vibration peaks of the corresponding C-N bonds are much sharper as shown in Figure 5a and 5b.

Vanadium Oxidation State

Since all these compounds decompose to pure V_2O_5 upon heating in oxygen, the V_2O_5 residue can be read directly from the TGA curve, and the amount of vanadium (V^{4+} and V^{5+}) in the starting material can be readily determined. The oxidation state of the vanadium was determined by the method described in the experimental section. The number of moles of V^{4+} is equal to that of consumed Ce^{4+}. The oxidation states of vanadium for $(CH_3NH_3)V_3O_7$, $(CH_3NH_3)_{0.75}V_4O_{10} \cdot 0.67H_2O$, and $(CH_3NH_2)_2V_8O_{17}$ are 4.34, 4.82, 4.26 respectively.

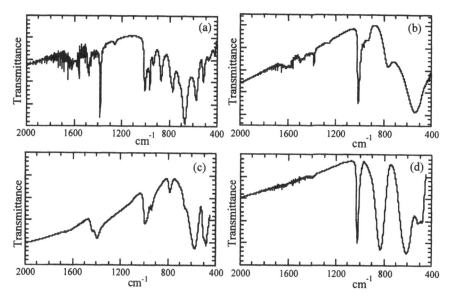

Fig. 5. The FTIR spectrum of (a) $(CH_3NH_3)V_3O_7$, (b) $(CH_3NH_3)_{0.75}V_4O_{10}\cdot0.67H_2O$, (c) $(CH_3NH_2)_2V_8O_{17}$, and (d) vanadium pentoxide, V_2O_5.

CONCLUSIONS

The utilization of the hydrothermal technique in combination with organic templates, such as tetramethylammonium and ethylenediamine, has resulted in the syntheses of a number of new vanadium oxides. In the case of tetramethylammonium, the structures have been determined for the $N(CH_3)_4V_4O_{10}$ [2, 4], $N(CH_3)_4V_3O_7$ [3], and $N(CH_3)_4V_8O_{20}$ [5] phases, formed over a wide range of pH using a combination of single crystal and powder Rietveld methods. This work has shown that methylamine also forms layered compounds with vanadium oxide, two of whose structures have been determined. One of these contains V_3O_7 chains as found in $N(CH_3)_4V_3O_7$, but the bonding between the square pyramidal chains and the tetrahedra differs. The second structure, formed under more acidic conditions contains double sheets of V_2O_5 layers as in the $\delta-Ag_xV_2O_5$ phase. The structure of a third phase is under investigation, and the IR data would suggest that it might contain a bond between the nitrogen and vanadium. These layered vanadium oxides were fully characterized, and their structures are readily reduced by the intercalation of lithium ions. We are presently studying other intercalation and ion exchange reactions with for example amines and hydrogen. Chemical or thermal removal of the organic templates from the lattices could lead to new vanadium oxides which may be of interest as catalysts and as the cathode in secondary lithium batteries.

ACKNOWLEDGMENT

We thank the Department of Energy through Lawrence Berkeley Laboratory and the National Science Foundation through grant DMR-9422667 for partial support of this work. We also thank Professor Jon Zubieta and Dave Rose at Syracuse University for the use of the Siemens Smart CCD single crystal diffractometer.

REFERENCES

1. M. S. Whittingham, J. Li, J. Guo, and P. Zavalij, Materials Science Forum, **152-153,** 99 (1994).
2. M. S. Whittingham, J. Guo, R. Chen, T. Chirayil, G. Janauer, and P. Zavalij, Solid State Ionics, **75,** 257 (1995).
3. T. A. Chirayil, P. Y. Zavalij, and M. S. Whittingham, Chem. Commun., 33 (1997).
4. P. Zavalij, M. S. Whittingham, E. A. Boylan, V. K. Pecharsky, and R. A. Jacobson, Z. Kryst., **211,** 464 (1996).
5. T. Chirayil, P. Y. Zavalij, and M. S. Whittingham, J. Mater. Chem., **7,** 2193 (1997).
6. F. Zhang, P. Y. Zavalij, and M. S. Whittingham, Mater. Res. Bull., **32,** 701 (1997).
7. R. C. Haushalter and L. A. Mundi, Chem. Mater., **4,** 31 (1992).
8. G. G. Janauer, P. Y. Zavalij, and M. S. Whittingham, Chem. Mater., **9,** 647 (1997).
9. K. P. Reis, A. Ramanan, and M. S. Whittingham, Chem. Mater., **2,** 219 (1990).
10. L. G. Akselrud, P. Y. Zavalii, Y. N. Grin, V. K. Pecharsky, B. Baumgartner, and E. Wolfel, Materials Science Forum, **133-136,** 335 (1993).
11. S. Andersson, Acta Chem. Scand., 1371 (1965).
12. Y. Oka, T. Yao, and N. Yamamoto, J. Mater. Chem., **5,** 1423 (1995).
13. J.-M. Savariault and J. Galy, J. Solid State Chem., **101,** 119 (1992).
14. Y. Kanke, K. Kato, E. Takayama-Muromachi, and M. Isobe, Acta Cryst., **C46,** 536 (1990).
15. T. Yao, Y. Oka, and N. Yamamoto, Mater. Res. Bull., **27,** 669 (1992).
16. L. F. Nazar, B. E. Koene, and J. F. Britten, Chem. Mater., **8,** 327 (1996).
17. Y. Zhang, R. C. Haushalter, and A. Clearfield, J. Chem. Soc., Chem. Commun., 1055 (1996).
18. Y. Zhang, R. D. DeBord, C. J. O'Connor, R. C. Haushalter, A. Clearfield, and J. Zubieta, Angew. Chem. Int. Ed. Engl., **35,** 989 (1996).
19. R. Chen, P. Y. Zavalij, and M. S. Whittingham, J. Mater. Chem., **8,** in press (1998).
20. P. Y. Zavalij, T. Chirayil, and M. S. Whittingham, Acta Cryst, **C53,** 879 (1997).

Part V
Zeolited and Related Materials

HYBRID CATALYSTS: INTERNAL OR EXTERNAL CONFIGURATION FOR BETTER CATALYTIC PERFORMANCE?

R. LE VAN MAO, M.A. SABERI, J.A. LAVIGNE, S. XIAO AND G. DENES Department of Chemistry and Biochemistry, Concordia University, 1455 De Maisonneuve West, Montreal, Quebec, H3G 1M8 Canada

ABSTRACT

Hybrid catalysts with the external configuration for the cocatalyst showed enhanced product diffusion rates in the n-octane hydrocracking, only if the reaction was carried out at relatively high temperatures. In the n-heptane isomerization, direct incorporation of the Al species into the HY zeolite micropores produced sorption sites which positively affected the selectivity to liquid isomers. Such as internal hybrid configuration resulted in a more important increase in the liquid isomer selectivity than that given by the external hybrid configuration, both systems being designed for better product outward-diffusion.

INTRODUCTION

Hybrid catalysts comprising micron-sized/microporous zeolite crystallites (main component) embedded in larger mesoporous particles (cocatalyst, assumed to be catalytically inert) showed enhanced product yields, when compared to the parent zeolite, in several reactions [1]: aromatization of light paraffins [2,3], selective hydrocracking of long-cabin paraffins (to produce isobutane and other isoparaffins) [4] and up-grading the products of steam-cracking of hydrocarbons [5]. Such enhanced performances were ascribed to the formation of a funnel-shaped pore continuum (micropore + mesopore) whose effect would be to decrease the energy barrier existing at the micropore opening of a zeolite by smoothing the change of the surface curvature [6]. Such an explanation is quite plausible because all the reported examples [1-5] related to catalytic reactions performed at relatively high temperatures: in fact, according to Barrer [7], the barrier of diffusion via an externally adsorbed layer as conventionally devised for zeolite crystals becomes less significant at high reaction temperatures.

The objectives of this research project were as follows: i) to perform the hydro-cracking/isomerization of n-octane over a wide temperature range by making use of the parent Pt-HY zeolite and the hybrid Pt-HY/SA (SA=silica-alumina used as cocatalyst). The Arrhenius plot resulting from these kinetic studies could allow us to determine at which reaction temperature the cocatalyst started having some effects on the diffusion of the products; and ii) by using the information from the n-octane reaction, to prepare hybrid catalysts for the hydroisomerization of the n-heptane. These hybrid catalysts would contain Al species incorporated as "diffusion accelerators" in two different ways: impregnation into the zeolite crystallites, or extrusion (embedment) with these crystallites, resulting in the "internal" and "external" hybrid configurations, respectively.

EXPERIMENTAL

I)   STUDY OF THE N-OCTANE HYDROCRACKING:
      i) Catalyst preparation and characterization:
Some of the physico-chemical properties of the zeolite HY (acid form, powder from U.O.P., Si/Al = 2.7) and the silica-alumina SA (Aldrich, grade 135, activated at 550°C for 10h) are reported in Table 1. Techniques for the catalyst characterization used in this study were described elsewhere [4, 8]. The procedure for the preparation of the final catalyst extrudates using betonite as binder has been described in ref. 4.

The Pt-HY and Pt-HY/SA both contained 80 wt % of HY zeolite and 0.5 wt % of Pt. Pt-HY/SA sample contained in addition 10 wt % of SA, being a hybrid catalyst with "external" configuration, i.e. the micro-sized zeolite crystallites being embedded in the larger-sized particles of silica-alumina. Another reference sample, Pt-SA (Pt = 0.5 wt %, SA = 80 wt % and balance = bentonite) was also prepared.

TABLE 1 Some physico-chemical properties of the HY zeolite and the cocatalyst used in the n-octane conversion

	Particle size (μm)[§]	BET surface area (m^2/g)	Pore diameter (10^{-9} m)	Surface Acidity[•] (mmol/g)	(mmol/m^2)
HY	1-2	610	0.74	0.92	1.5
SA	50-100	641	4.6	0.32[†]	0.5

([§]) determined by scanning electron microscopy;
([•]) measured by the NH_3 TPD technique;
([†]) weak acid sites when compared with those of the HY zeolite.

ii) Catalyst testing and data computing:
a) The experimental set-up for the conversion of n-octane was identical to that described elsewhere [4, 8]. The reaction parameters were as follows: temperature = 172 °C-297 °C (445 K - 570 K); contact time = 0.13 to 2.5 h; partial pressure of n-octane = 0.02 atm; ratio of flow-rate of nitrogen (used as carrier gas) to flow-rate of hydrogen = 4/3; weight of catalyst = 0.625 g and duration of a run = 4h. The testing procedure and product analysis techniques used were identical to those described elsewhere [8, 9].
b) The total conversion of n-octane is defined as follows:

$$C_t(C\ atom\%) = 100\frac{[(NC)_f - (NC)_P]}{(NC)_f}$$

where $(NC)_f$ and $(NC)_P$ are the numbers of C atoms of n-octane in the feed and in the reactor outstream, respectively.
For each catalyst at each reaction temperature, the experimental data for total conversion (much lower than 30 C atom %) were fit to a function, $f(\tau)$, of the contact time τ, by using a non-linear regression algorithm. A third-order polynomial function was found with a good correlation factor (>0.98): $C_t = a\tau + b\tau^2 + c\tau^3$ at a given reaction temperature T. The $r = kP^n$ general form of the rate equation was used, where k, P and n were the rate constant, the partial pressure of n-octane and the order of the reaction, respectively. For accuracy reason, data of initial rates were used, resulting in the following equation for a reaction temperature T:

$$K = P^{-n} \times r_0 = P^{-n}\left[\frac{df}{d\tau}\right]_{\tau=0} = C \times r_0$$

where C is a constant and r_0 is the initial rate of reaction. From these data, Arrhenius plots ($1n\ r_0$ versus $10^3/T$) were obtained.

II) STUDY OF THE N-HEPTANE ISOMERIZATION:

i) <u>Catalyst preparation and characterization:</u>

The HY materials were obtained by calcination in air at 550 °C overnight of the ammonium-Y zeolites [LZY-84 provided by U.O.P. in two forms: 1) 1/16" pellets (containing ca. 80 wt % of NH_4Y and some non-defined large pore-sized binder), and 2) fine powder of NH_4Y].

a) $Pt-Al_{iy}$ -HY (i stands for "internal"):

10 g of HY zeolite pellets were immersed in a solution of x g of Al $(NO_3)_3$. $9H_2O$ (Aldrich, ACS reagent) dissolved in 15 ml of water. x was calculated so that the Al content of the hybrid catalyst was equal to y, y varying from 1 to 7 wt %. After evaporation of the suspension to dryness on a hot plate, the solid was dried at 120 °C overnight and then subsequently activated in air at 350 °C for 10 h. The resulting solid was immersed in a solution of 0.172 g of Pt (II) tetramine chloride hydrate dissolved in 15 ml of water. After evaporation of the suspension to dryness on a hot plate, the solid was dried at 120 °C overnight and then activated in air at 350 °C for 10 h. These hybrid catalysts are assumed to have Al species (finally decomposed to Al_2O_3) incorporated directly into the micropores of the HY zeolite (internal hybrid configuration).

b) $Pt-Al_{ey}$-HY (e stands for "external"):

Al_2O_3 used to prepare the hybrid catalyst was obtained by calcining in air the $Al(NO_3)_3$. $9 H_2O$ at 350 °C overnight. HY powder was treated with a solution of Pt (II) tetramine chloride hydrate in the same way as for the $Pt-Al_{iy}$-HY system. A mixture of the resulting Pt-HY zeolite powder (80 wt %), the alumina previously produced (y wt %) and bentonite (binder, balance) was used for the preparation of the extrudates of the hybrid catalyst (external configuration), following the procedure previously described elsewhere [4, 8].

c) $(Pt-HY)_i$ and $(Pt-HY)_e$ (reference samples):

1) $(Pt-HY)_i$: 10 g of the HY pellets were immersed in a solution of 0.172 g of Pt(II) tetramine chloride hydrate dissolved in 15 ml of distilled water. After evaporation of the suspension to dryness on a hot plate, the zeolite material was dried at 120 °C overnight and activated in air at 350 °C for 10 h.

2) $(Pt-HY)_e$: same procedure as with the $Pt-Al_{ey}$-HY, thus using the HY powder but without any Al_2O_3.

The techniques used for the characterization of the catalysts (in extrudate form) were identical to those described elsewhere [4, 8].

TABLE 2 Physico-chemical properties of the catalysts studied

Catalyst	Al content (wt %)	BET surface area (m²/g)		Micropore size nm
		total	micropores	
$(Pt-HY)_i$	0	506	372	0.74
$Pt-Al_{i1}$-HY	1	444	315	0.74
$Pt-Al_{i2}$-HY	2	437	310	0.73
$Pt-Al_{i5}$-HY	5	439	319	0.73
$Pt-Al_{i7}$-HY	7	363	265	0.72
$(Pt-HY)_e$	0	514	425	0.74
$Pt-Al_{e2}$-HY	2	517	431	0.74

ii) Catalyst testing:

a) The experimental set-up for the n-heptane isomerization was identical to that described elsewhere [4, 8]. The reaction parameters were as follows: temperature = 225°C (498 K); contact time = 1.1 h; flow-rate of hydrogen = 15.4 ml/min; weight of catalyst = 2.0 g; weight hourly space velocity (g of n-heptane injected per hour per gram of catalyst) = 0.94 h^{-1}; and duration of a run = 2.5 h. Prior to the testing, the catalyst was reduced in-situ under hydrogen atmosphere at 275°C for 2 hours (Pt content = 1 wt %). The testing procedure and product analysis techniques used were identical to those described elsewhere [4, 8].

b) The total conversion of n-heptane was defined as for the conversion of n-octane. The selectively and yield of product Pi were given, respectively, by:

$Y_{pi} = 100 \times (NC)_{Pi} / (NC)_f$ and $S^{Pi} = Y_{Pi} / (C_t \times 100)$, where $(NC)_{Pi}$ is the number of C atoms of product Pi.

RESULTS AND DISCUSSION
I) N-OCTANE CONVERSION:

Figure 1 shows the Arrhenius plots of the Pt-HY parent zeolite (a) and the Pt-HY/SA hybrid catalyst (b). While the slope of the curve 1a (parent zeolite) is the same for all the points corresponding to the temperature range investigated, there is a clear slope change for curve 1b (hybrid catalyst) starting at a temperature of 225 °C (498 K). This is indicative of the occurrence of a diffusional phenomenon leading to a lower value of the apparent activation energy. Since the reference sample Pt/SA (silica-alumina alone) did not give any significant n-octane conversion under the same reaction conditions, the effect of the cocatalyst could not be attributed to any cooperation of the cocatalyst surface to the total conversion. The mesoporosity and the large surface area of the cocatalyst used, led to the same interpretation as in our hypothesis where the formation of a pore continuum within the hybrid catalyst was proposed. In fact, the beneficial effect of the newly formed funnel-shaped pore system on the outward-diffusion of the reaction products appeared only with reactions occurring on high temperature [1].

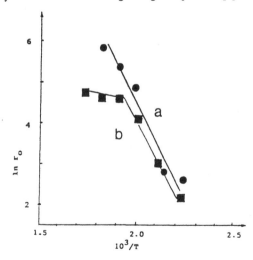

FIGURE 1. Arrhenius plots obtained with the Pt-HY zeolite (a) and the hybrid catalyst, Pt-HY/SA (b)

In the conversion of n-octane, therefore, the catalyst with an external hybrid configuration behaved like the parent zeolite when the reaction temperature was lower than 225 °C (498 K).

II) N-HEPTANE ISOMERIZATION:

The results obtained with the hybrid catalysts having the external configuration for the cocatalyst (Al_2O_3), tested in the isomerization of n-heptane at 225 °C, are reported in Table 3. There was significant increases of the selectively (S_{Liso}) and the yield (Y_{Liso}) of liquid isomers (C_5-C_7) for the hybrid catalyst (Pt-Al_{e2}-HY) when compared to the parent zeolite, (Pt-HY)$_e$. This indicates that the reaction temperature used was sufficiently high to promote the interaction between the two components of the Pt-Al_{e2}-HY sample.

TABLE 3. Performances of some catalysts investigated in the isomerization of n-heptane

Catalyst	Total conversion (C_t in wt%)	Y_{Liso} (wt%)	mu/mo[*]	iso/cr[§]	S_{Liso} (wt%)
(Pt-HY)$_e$[a]	76.1	45.8	0.31	0.9	51
Pt-Al_{e2}-HY	83.6	53.6	0.35	1.1	64
(Pt-HY)$_i$[b]	69.8	40.8	0.30	0.9	58
Pt-Al_{i1}-HY	67.4	53.6	0.31	1.1	80
Pt-$Al_{i1.5}$-HY	72.5	44.0	0.32	1.2	62
Pt-Al_{i2}-HY	75.1	45.1	0.31	1.2	60
Pt-$Al_{i2.5}$-HY	65.2	40.8	0.28	1.3	63
Pt-Al_{i3}-HY	64.3	36.9	0.27	1.1	57
Pt-Al_{i5}-HY	58.6	36.6	0.34	1.2	62

(*) multibranched/monobranched; (§) isomerization/cracking;
(a) external configuration series; (b) internal configuration series.

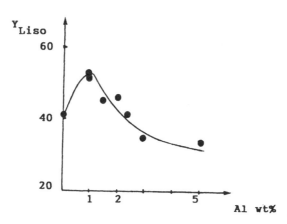

FIGURE 2 Influence of the Al loading on the yield of liquid isomers

With the hybrid catalysts prepared by direct incorporation of Al species (Pt-Al$_{iy}$-HY samples), the yield of liquid isomers went through a maximum at ca. 1.0 wt % of Al (Figure 2 and Table 3). Since at such an Al concentration: (a) the pore size of the hybrid catalyst did not change significantly (Table 2), (b) the selectivity to liquid isomers S_{Liso} was significantly higher than that of the parent zeolite (Table 3), (c) the total and micropore surface areas decreased significantly with the Al loading (Table 2), the increase of the yield of liquid isomers could be ascribed to new sorption sites formed by the incorporated Al species located next to the zeolite acid sites. Such Al sites are capable of shortening the residence time of the reaction intermediates on the acid sites, thus slowing down the cracking of branched intermediates with the multibranched carbocations being the most sensitive to cracking.

Moreover, the enhancement of the selectivity, S_{Liso}, and the yield, Y_{Liso}, obtained with the internal hybrid configuration (when compared to the corresponding parent zeolite) was more important than that obtained with the external hybrid configuration. Therefore, both hybrid systems involved the same physical phenomenon: a more rapid evacuation of products or reaction intermediates.

CONCLUSION

Direct incorporation of Al species into the large pore-sized Y zeolite resulted in enhanced selectivity to liquid isomers in the isomerization of long-chain paraffins. With such an internal hybrid configuration, reaction intermediates were rapidly evacuated by the newly formed adsorption sites, thus preventing excessive cracking. The ratio of multibranched/monobranched products was kept high. This is in contrast with the present general approach which consists of using medium pore sized zeolite materials [9-11] as highly selective catalysts: the production of liquid isomers is actually high, however, these are almost monomers that are known to be poor octane rating compounds (for transportation fuels) when compared to the multibranched hydrocarbons.

ACKNOWLEDGMENTS

We would like to thank NSERC of Canada for financial support, Mr. S.T. Le for technical assistance and U.O.P. (Desplaines, IL, USA) for having kindly provided us with the HY zeolites.

REFERENCES

1.   R. Le Van Mao, Zeolites, submitted.
2.   J. Yao and R. Le Van Mao, Catal. Lett. **11**, 191 (1991).
3.   R. Le Van Mao, J. Yao, L. Dufresne, R. Carli, Catal. Today **31**, 247 (1996).
4.   S. Xiao and R. Le Van Mao, Microporous Mater. **4**, 435 (1995).
5.   R. Le Van Mao, U.S. Patent 4,732,881 (Mar. 22, 1988).
6.   E.G. Derouane, J.M. Andre and A.A. Lucas, J. Catal. 110, 58 (1988).
7.   R.M. Barrer, J. Chem. Soc., Faraday Trans. **86** (7), 1123 (1990).
8.   S. Xiao, R. Le Van Mao and G. Denes, J. Mater. Chem. **5** (6), 1251 (1995).
9.   S.J. Miller, Microporous Mater., **2**, 439 (1994).
10. R.J. Taylor and R.H. Petty, Applied Catal. A, **119**, 121 (1994).
11. B. Parlitz, E. Schreier, H.L. Zubowa, R. Eckelt, E. Lieske, G. Lischke and R. Fricke, J. Catal. **155**, 1 (1995).

Part VI

Acid and Bases

MIXED METAL PHOSPHO-SULFATES FOR ACID CATALYSIS

S. G. Thoma*, N. B. Jackson*, T. M. Nenoff*, R. S. Maxwell**
*Sandia National Laboratories, Albuquerque, NM 87185-0710
**Lawrence Livermore National Laboratory, Livermore, CA 94551

ABSTRACT

Mixed Metal Phospho-Sulfates were prepared and evaluated for use as acid catalysts via 2-methyl-2-pentene isomerization and o-xylene isomerization. Particular members of this class of materials exhibit greater levels of activity than sulfated zirconia as well as lower rates and magnitudes of deactivation. ^{31}P MAS NMR has been used to examine the role of phosphorous in contributing to the activity and deactivation behavior of these materials, while powder X-ray diffraction, BET surface area, IR, and elemental analysis were used to characterize the bulk catalysts.

INTRODUCTION

The use of metal phosphates and sulfated metal oxides as solid acid catalysts has been the subject of much study, and the chemical mechanism responsible for acid properties in solids the subject of much debate. Historically, Lewis acidity in metal phosphates has been attributed to dehydroxylated surface metal sites, and Bronsted acidity to hydrated Lewis acid sites where the inductive effect of neighboring non-hydrated Lewis acid sites increases the lability of protons in the adsorbed water molecule[1]. Surface P-OH groups were found to be the source of Lewis and Bronsted acid sites by others[2,3]. Segawa, et. al.[4,5,6] suggested that it is the electron withdrawing capability of bulk phosphate groups that enhance the acid strength of surface P-OH groups, with the electrons being withdrawn into the bulk via P-O-P bonds.

Ward & Ko[7,8] described a similar mechanism to account for Bronsted acidity in sulfated zirconia. They suggest that the inductive power of a sulfate group, due to the strong electronegativity of the S=O bonds, promotes the lability of protons on an adjacent hydrated surface metal site. More recently, Farcasiu, et. al.[9] and Ghenciu, et. al.[10] have suggested that the activity typically attributed to Bronsted acidity in sulfated metal oxides is due to a reduction-oxidation mechanism. They showed that sulfate groups act as a one-electron oxidizing agent, which in conjunction with neighboring hydroxyl groups, initiate and propagate chemical reactions. This concept has been further supported by Xu and Sachtler[11] who correlated the activity associated with Bronsted acidity on sulfated zirconia with the oxidation state of the sulfur atom.

In this study a series of metal and mixed-metal phospho-sulfates have been prepared which contain greater amounts of bulk and surface sulfate groups than is typically found in solid acid catalysts. These materials have reasonably high surface areas and a minimum of 10 wt% sulfur which is completely stable in oxidizing environments to 450 °C. The structure of these materials is explored using [31]P Magic-Angle Spinning Nuclear Magnetic Resonance (MAS NMR), X-ray powder diffraction (XRD), and Fourier Transform Infra-Red Spectroscopy (FTIR). An attempt is made to correlate structure with catalytic activity as observed by 2-methyl-2-pentene and o-xylene isomerizations.

EXPERIMENTAL

Synthesis & Characterization

Catalysts were synthesized by reacting mixed-metal solutions with stoichiometric amounts of phosphoric acid and water, and an excess of sulfuric acid. The resultant precipitates were heated under flowing air at a ramp rate of 0.1 °C /minute to 300 °C and held at this temperature for 72 hours. The calcined materials were ground with a mortar and pestle afterwhich they were granulated by alternating 2000 PSI hydraulic loading, grinding, and sieving resulting in sample particles of 100-200 μm.

Powder X-ray diffraction data were collected at room temperature on a Siemens Model D500 diffractometer, with Θ-2Θ sample geometry and Cu Kα radiation, between 2Θ = 5 and 60°. BET measurements were performed on a Quantachrome Autosorb automated gas sorption system, with adsorbed and desorbed volumes of nitrogen in the relative pressure range of 0.05-1.0. Chemical analysis was performed via DCP using an ARL SS-7 DCP, with the exception of sulfur analysis, which was performed by Galbraith Laboratories. ^{31}P MAS NMR data were collected on a 9.4T system at 400 MHz for protons and 162 MHz for phosphorous, with a 90° pulse and spinning speeds between 6-7 KHz. The chemical shifts are relative to phosphoric acid as an external standard. De-convolution of spectra was performed using Peakfit 4.0 software (Jandel).

The FTIR spectra of the samples were obtained using both transmission and diffuse reflectance modes. The transmission infrared spectroscopy was obtained with a Mattson Research Series 1 FTIR spectrometer equipped with an air-cooled source and a mercury-cadmium-telluride detector. The spectra were obtained at a resolution of 4 cm^{-1} after 100-200 scans. The diffuse reflectance spectra were obtained using a Nicolet 20SXB FTIR spectrometer equipped with a water-cooled source and a mercury-cadmium-telluride detector. The spectra were obtained at a resolution of 4 cm^{-1} after 250 scans. The IR cell was a Spectratech diffuse reflectance assembly. All spectra were taken at ambient temperature and in the N$_2$ purge of the IR chamber. The samples were diluted to a 5% sample in KBr mixture.

Isomerization

The materials were tested for activity toward 2-methyl-2-pentene isomerization and o-xylene isomerization using a packed bed flow reactor. Approximately 50 mg of granulated sample was placed in a 6 mm O.D., 40 cm long fritted glass tube and held in place with glass wool. The samples were pretreated at elevated temperatures in ultra-high purity helium (99.999% Matheson) flowing at 40-60 cc/min. All pre-treatments and reactions were at atmospheric pressure.

After pretreatment, the catalyst sample was cooled and the reaction was run isothermally at 150 °C. Flowing He was used to carry the 2-methyl-2-pentene and o-xylene from the 0 °C saturator to the reactor bed, at 10 cc/min and 1.5-2.0 cc/min, respectively. The products and reactants flowed through heated lines to a HP 5890 series II gas chromatograph (GC) equipped with a flame ionization detector (FID). The GC was equipped with an automatic sampling valve that injected 1 cc of reactor effluent into a Supelco SPB-1 column for pentenes or a Supelco Alphadex column for xylenes, connected to the FID.

RESULTS & DISCUSSION

Characterization Results

Table 1 lists the results of BET surface area and XRD analysis for samples used for catalytic testing. In general, the metal to phosphorous ratio was fixed at 1.3:1.0, and in the mixed metal systems the ratio of metal 1:metal 2 was 1.0:1.6. (Zr/Ti/P/S has a molar Zr:Ti:P ratio of 1.0:1.6:2.0, etc). The samples that contained sulfur were 10-15 wt% sulfur. XRD analysis showed that the samples which exhibited crystallinity were in the NASICON system with structures of the type $M1_xM2_{(2-x)}(PO_4)_2(SO_4)$. Based upon the synthesis method employed and the precursor stoichiometry, the tendency to crystallize in the NASICON system was not unexpected[12]. However, the XRD peaks are very broad and set in the low humps generally associated with amorphous materials, suggesting that the material is not fully crystalline. The lack of complete crystallinity may be due to the lack of a balancing counter-ion or precursor non-stoichiometry with respect to $M_2(PO_4)_2(SO_4)$.

TABLE 1

Phospho-Sulfate Catalyst BET Surface Area and XRD Results

Sample	(m^2/g)	XRD
Zr/P/S	4	NASICON/amorphous
Zr/Ti/P	434	amorphous
Zr/Ti/P/S	35	NASICON/amorphous
Zr/Al/P/S	47	NASICON/amorphous
Zr/Ta/P/S	41	NASICON/amorphous
Ta/P/S	60	amorphous

Structural calculations were performed using a formal charge model that assumes the most stable configuration for a material is when the net charge on the greatest number of oxygen atoms is minimized. This model has been used to predict the structure and relative acidity in mixed-metal oxides[13] but was extended here to include metal phosphates and sulfates. Test calculations were performed on naturally occurring metal phosphate and metal sulfate minerals as a means to validate the incorporation of these additional elements. The stoichiometry of the catalysts as determined by elemental analysis was used as the basis of calculation in the formal charge model. The most stable configuration was determined to be the $M_2(PO_4)_2(SO_4)$ NASICON structure. Two other equally favorable configurations have increased numbers of M-O-S bonds at the expense of M-O-P bonds *which correspondingly increased the numbers of non-bonding oxygen atoms (NBO's) in the form of hydroxyl groups.* The increase in NBO's creates a situation where the material can exist with a consistent stoichiometric yet non-systematic structure (i.e. amorphous). We therefore suggest that the structure of the catalysts is comprised of a continuum of these NASICON and amorphous structural elements. The local interaction between sulfate groups and the metals determines the relative amounts of NASICON type crystal structure or an equally stable amorphous phase.

^{31}P MAS NMR Results

^{31}P MAS NMR of the six phospho-sulfate catalysts yielded a set of three different groups of spectra based on similarity of the overall shape and chemical shifts observed. The sets are labeled Groups 1-3 and each set had two samples associated with it. An example from each of these groups is presented in Figure 1. In general, the spectra are characterized by relatively broad and highly overlapping resonances that limited spectral resolution and decreased the precision of the de-convolutions. The spectral de-convolution results are given in Table 2. The large range of chemical shifts in each group is explained by distributions in level of hydration, in hydrogen bonding strengths, in phosphorous next nearest neighbors, and in distortions to the tetrahedral units, all of which are known to affect ^{31}P chemical shifts. Further experiments need to be conducted in order to make conclusive peak assignments, but the general trend is that the more negative peaks most likely correspond to more polymerized phosphate networks while more positive peaks correspond to an increase in NBOs terminated by protons [4,5,14,15]. The final category (<-40 ppm) may include contributions due to condensed P-O-P units, however, the presence of these groups is unexpected.

FIGURE 1

^{31}P MAS NMR Results

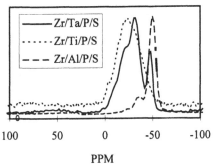

PPM

TABLE 2

^{31}P NMR Results

Notation = chemical shift (% total area)

Group 1		Group 2		Group 3	
Zr/Ti/P	Zr/Ti/P/S	Zr/P/S	Zr/Al/P/S	Zr/Ta/P/S	Ta/P/S
	-52.3 (3%)	-47.2 (67%)	-49.2 (84%)	-47.3 (9%)	-46.0 (1%)
-36.1 (21%)					
-31.3 (20%)	-31.5 (26%)	-34.3 (28%)	-35.3 (13%)	-31.1 (71%)	-31.4 (34%)
-17.1 (46%)	-19.6 (44%)	-23.6 (5%)	-23.1 (3%)	-15.9 (10%)	-21.6 (43%)
-5.7 (4%)	-11.0 (15%)			-11.1 (9%)	-11.6 (22%)
-0.9 (4%)	-0.4 (12%)				
+2.4 (4%)				+2.2 (1%)	

The high degree of peak overlap characteristic of the Group 1 samples (Table 2) complicated de-convolution. The general trend, however, shows the tendency of sulfur to inhibit metal-phosphate group condensation, demonstrated by the overall increase in phosphorous with NBO's in the Zr/Ti/P/S versus the Zr/Ti/P.

The Group 2 samples have a denser structure (with respect to phosphorous) than Group 1 samples, in which the majority of phosphate groups are fully coordinated. The fact that the aluminum containing sample is structurally similar to the pure zirconium sample suggests that the presence of aluminum allowed for formation of a more condensed structure. If the Zr/Al/P/S is denser than the Zr/Ti/P/S of group 1, the number of P-O-M bonds has increased at the expense of phosphorous NBO's.

Group 3 includes both of the tantalum containing samples, which consist largely of phosphate tetrahedra with 3 BOs. The fact that the Zr/Ta/P/S has a significant -50ppm shift as compared with the Ta/P/S suggests that this shift may be attributed to zirconium-phosphate coordination. Also, since the NASICON structure implies octahedral coordination of metals and the Ta/P/S has a dissimilar spectra to Zr/P/S we conclude that tantalum is not octahedrally coordinated. This is supported by XRD results which show that decreasing the Zr:Ta molar ratio gradually decreases the diffraction peak intensities until there is no crystalline phase present at all, (as is the case with the Ta/P/S). Furthermore, NASICON materials of the type $Ta_xM_{(2-x)}(PO_4)_3$ (M= Zr or Ti) are known to exist but none of the type $Ta_xM_{(2-x)}(PO_4)_2(SO_4)$[12].

2-methyl-2-pentene Isomerization Results

2-methyl-2-pentene isomerization is an acid catalyzed reaction in which distribution of the three major products yields information about the relative strength of acid sites on the catalyst[16,17,18]. The rearrangement of 2-methyl-2-pentene (2M2P) progresses by formal protonation of the olefin followed by isomerization and deprotonation of reaction intermediates[16]. *Weak* acid sites produce a double bond migration yielding 4-methyl-2-pentene (4M2P), *medium* strength acid sites produce a methyl shift yielding 3-methyl-2-pentene (3M2P), and *strong* acid sites produce an extensive rearrangement to 2,3 dimethylbutene (2,3DMB)[18]. The weak and medium sites have acid strengths that are typically attributed to Lewis acidity and the strong sites to Bronsted acidity.

The 3M2P production rate versus time is presented in Figure 2. Data for the strong and weak sites display the same relative position of samples to one another and similar rate versus time curve shape. Detailed 2M2P isomerization reaction data for these samples, presented elsewhere[19], has shown that catalytic deactivation under 450 °C was caused solely by carbon build-up, and the activity could be fully regenerated by subsequent treatment with oxygen. In general, the tantalum containing materials maintained a greater amount of their activity (over any time span). This resistance to deactivation was attributed to the ability of tantalum to transport oxygen from its matrix, preventing the deposition of carbon and hence, deactivation. Over 450 °C, deactivation is accompanied by sulfur loss. There is still carbon build-up but the activity is no longer fully regenerable[20].

2M2P isomerization results show that materials that are similar in structure (as monitored by ^{31}P MAS NMR) can have drastically different activities. For instance, the tantalum containing materials and the Zr/Al/P/S show significantly higher isomerization rates over the other samples, yet they have distinctly different NMR spectra. Furthermore, the

Zr/Al/P/S and Zr/P/S have nearly identical NMR spectra, yet Zr/Al/P/S is one of the most active samples and Zr/P/S is the least active. Thus, the state of phosphorous is not solely responsible for the observed activity.

FIGURE 2

Rate of 2M2P Isomerization (to 3M2P) versus Time

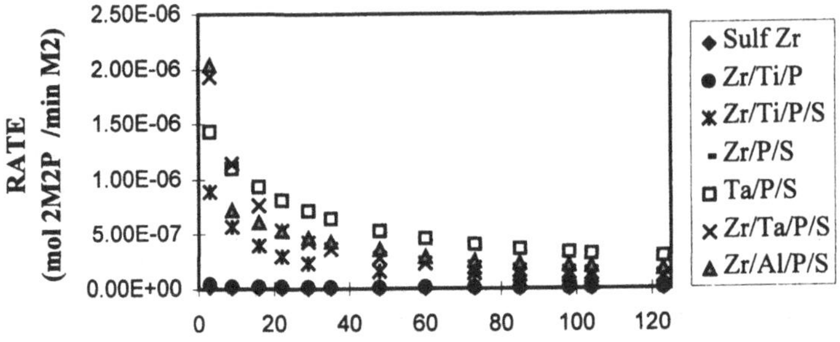

O-xylene Isomerization Results

The products from catalytic transformation of o-xylene are indicative of the catalytic mechanism. Three mechanisms have been identified as active in this reaction: 1) acid mechanism which yields isomerization products p- and m- xylene as well as disproportionation products toluene and benzene; 2) hydrogenation mechanism which yields dimethylcyclohexanes; 3) bifunctional mechanism (combined acid and hydrogenation) which yields trimethylcyclopentanes. Some isomerization and disproportionation products are produced from bifunctional catalysis but the amount is considered to be negligable[20].

O-xylene isomerization was performed over zeolite ZSM-5, sulfated zirconia, zirconium pyrophosphate, as well as the six phosphates and phospho-sulfates under consideration here. In general, ZSM-5 showed excellent conversion at lower temperatures and good selectivity toward p-xylene. Sulfated zirconia showed good conversion at higher temperatures but with less conversion to p-xylene. The zirconium pyrophosphate, mixed-metal phosphates, and phospho-sulfates on the other hand, showed no o-xylene isomerization activity.

Discussion

Since the 2M2P isomerization rate as a function of time is virtually the same, and the relative magnitudes of the 2M2P isomerization rate remain constant, for each sample and in each of the three acid site strength categories, the same catalytic mechanism is probably responsible for each isomerization product. This behavior, (combined with the relationship between regenerable activity and sulfur loss found with the phospho-sulfates) suggests that a mechanism consistent with that proposed by Farcasiu[10] is responsible for the activity rather than a purely acid type mechanism.

Differences between the activity of the phospho-sulfates and sulfated zirconia may be attributed to two sources, the different state of the sulfate groups and different type hydroxyl sites. Bonding in sulfated metal oxides between an adsorbed sulfate and the surface is less covalent in nature than the structurally incorporated sulfate groups. The decreased covalent character of the metal-sulfate bonds would thus tend to increase their oxidizing strength versus the structurally incorporated sulfate groups of the phospho-sulfate materials. Only the adsorbed sulfate can isomerize xylene, which has a higher activation energy toward isomerization than does 2M2P. Differences in activity between these two materials may also be due to the nature of the hydroxyl group. Sulfated zirconia has exclusively Zr-OH hydroxyls (bridging or coordinative), whereas IR analysis of the phospho-sulfates shows that at temperatures greater than 200 °C the metal sites were largely dehydroxylated and that only P-OH groups were still present at reaction temperatures.

The enhanced rate of 2M2P isomerization, as well as diminished deactivation behavior of the phospho-sulfates versus sulfated zirconia may be attributed to the greater sulfate content of the phospho-sulfates. Beyond the possibility of increased number of surface sulfate sites, structurally incorporated sulfate is stabilized against reduction to inactive sulfite by the covalency of the sulfate-metal bonding, and can continue to act as an oxidant by electron withdrawal from the surface to bulk sulfate groups. Deactivation occurs when enough carbon adsorbs on surface to block hydride transfer.

The reason why some metals show greater activity in mixed metal systems and others show greater activity in a single metal environment in the phospho-sulfate catalysts is still unclear. Physical differences in metal-sulfate interaction may play a role, such as bridging versus chelating bidentate sulfate bonding. It is also conceivable that changes in metal coordination or extent of metal polyhedra distortion can effect the strength of sulfate group oxidizing ability, although it is doubtful that this phenomena is independent of other factors. Catalytic activity has been shown to scale with metal valency, mixed-metal coordination[12], as well as metal electronegativity[21]. These results suggest that a relationship exists between the fundamental electro-chemical properties of the metal and catalytic activity. If the observed activity in phosphate, phospho-sulfate, and sulfated-metal oxide catalysts is due to an electron withdrawing mechanism, it should be possible to find an (at least partial) explanation in molecular orbital theory.

CONCLUSION

The catalytic activity in metal phospho-sulfates is largely a function of the electronic properties of the metals and the type and extent of interaction between the metal and the sulfate groups. Phosphate groups contribute to the activity by providing surface hydroxyls and stabilizing the metal-sulfate structure, as monitored by ^{31}P MAS NMR, XRD, FTIR, and elemental chemical analysis.

Work currently in progress will hopefully further our understanding between the fundamental electrochemical properties of metals, their interaction with sulfate groups, and the effect on catalytic activity. We believe that by gaining a better understanding of these issues that we will be able to design and synthesize more versatile and selective catalysts.

ACKNOWLEDGEMENTS

This work was supported by the United States Department of Energy under Contract DE-AC04-94AL85000. Sandia is a multiprogram laboratory operated by Sandia Corporation, a Lockheed Martin Company, for the United States Department of Energy. The authors would like to thank Susan Crawford of Los Alamos National Laboratories for collecting NMR data.

REFERENCES

1. K. Tanabe, Solid Acids and Bases, Kodansha, Tokyo, 1970.

2. T. Hattori, A. Ishiguruo, and Y. Murakami, J. Inorg. Nucl. Chem., **40**, p. 1107-1111 (1978).

3. A. Clearfield and D. S. Thakur, J. Catal., **65**, p. 185 (1980).

4. K. Segawa, Y. Nakajima, S. Nakata, S. Asaoka, and H. Takahashi, Journal of Catalysis, **101**, p. 81-89 (1986).

5. K. Segawa, S. Nakata, and S. Asaoka, Materials Chemistry and Physics, **17**, p. 181-200 (1987).

6. K. Segawa, Y. Kurushu, Y. Nakajima, and M. Kinoshita, Journal of Catalysis, **94**, p. 491-500 (1985).

7. D. A. Ward and E. I. Ko, Journal of Catalysis, **157**, p. 321-333 (1995).

8. D. A. Ward and E. I. Ko, Journal of Catalysis, **150**, p. 18-33 (1994).

9. D. Facasiu, A. Ghenciu, and J. Q. Li, Journal of Catalysis, **158**, p. 116-127 (1996).

10. A. Ghenciu and D. Farcasiu, Journal of Molecular Catalysis A: Chemical, **109**, p. 273-283 (1996).

11. B. Q. Xu and W. M. H. Sachtler, Journal of Catalysis, **167**, p. 224-233 (1997)

12. J. Alamo, Solid State Ionics, **63-65**, p. 547-561 (1993).

13. B. Bunker, Mat. Res. Soc. Symp. Proc. **432**, p. 20-25 (1997).

14. R. K. Brow, C. C. Phifer, G. L. Turner, R. J. Kirkpatrick, SAND90-0710J (1990).

15. S. Prabhakar, K. J. Rao, and C. N. R. Rao, Chemical Physics Letters, **139** (1), p. 96-102 (1987).

16. G. M. Kramer, G.B McVicker, and J. J. Ziemiak, J. Catal., **92**, p. 355-363 (1985).

17. G. M. Kramer and G. B. McVicker, Acc. Chem. Res., **19**, p. 78-84 (1986).

18. J. F. Brody, J. W. Johnson, G. B. McVicker, and J. J. Ziemiak, Solid State Ionics, **32/33**, p. 350-353 (1989).

19. N. B. Jackson, T. M. Nenoff, S. G. Thoma, and S. D. Kohler, SAND97-2394, 1997

20. M. Guisnet, C. Thomazeau, J. L. Lemberton, and S. Mignard, Journal of Catalysis, **151**, p. 102-110 (1995).

21. A. La Ginestra and P. Patrono, Materials Chemistry and Physics, **17**, p. 161-179 (1987).

DISPERSED COMPLEX PHOSPHATES AS TAILORED ACID CATALYSTS: SYNTHESIS AND PROPERTIES

Vladislav A. Sadykov*, D.I. Kochubei*, S.P. Degtyarev*, E.A. Paukshtis*, E.B. Burgina*, S.N. Pavlova*, R.I. Maximovskaya*, G.S. Litvak*, V.I. Zaikovskii*, S.V. Tsybulya, V.F. Tretyakov**, D. Agrawal***, R. Roy***.
*Boreskov Institute of Catalysis SD RAS, Novosibirsk, Russia, 630090, sadykov@catalysis.nsk.su
**Topchiev Institute of Petrochemical Synthesis RAS, Moscow, Russia.
***Materials Research Lab, University Park, PA.

ABSTRACT

The relationship between the chemical composition, bulk real structure, surface properties and catalytic activity of dispersed complex framework zirconium phosphates with a NASICON-type structure is considered. For both crystalline and amorphous samples, a model of their bulk structure is suggested. The Lewis and Brönsted acidity was shown to vary broadly and be dependent upon the composition and preparation procedure, which is important for catalysts of hydro-carbons activation. Selective catalytic reduction of NO_x by methane in excess oxygen appears to be one of the application areas for these systems.

INTRODUCTION

Recently, sulfated zirconias and layered zirconium phosphates have attracted a great deal of attention as low-temperature acid catalysts for the processes of hydrocarbons transformation (hydrocracking, isomerization etc.) Unfortunately, they suffer from premature deactivation due to a loss of sulfates or recrystallization [1]. Framework zirconium phosphates with NZP or NASICON-type structure appear to be promising candidates for catalytic application due to their stability, structure flexibility enabling one to incorporate various substituting cations into the lattice [2], and, hence, to regulate their acid/base and/or oxidation/reduction properties. Up to date, only one study has been published concerning properties of vanadium-containing framework phosphates in the reactions of selective oxidation [3]. In part, it was explained by a lack of data concerning properties of dispersed low-temperature framework phosphates as dependent upon the composition and preparation conditions. The major objective of this is to elucidate the main features of the bulk and surface structure of highly dispersed amorphous/crystalline framework zirconium phosphates and gain an understanding of possible limits of the variations of their acid properties. The impact of acidity on catalysis of the NO_x selective reduction by hydrocarbons known to the catalyzed by such typical acid catalysts as H-ZSM-5 is also considered.

EXPERIMENTAL

Samples with nominal compositions $MeZr_4P_6O_{24}nH_2O$ (Me-Ca, Sr, Co, Mn, Fe or their mixture), $Me_{2/3}Zr_4P_6O_{24}nH_2O$ (Me=La) and $Zr_{0.25}Zr_4P_6O_{24}nH_2O$ were synthesized as in [4] by adding phosphoric acid to solution of corresponding nitrates. The precipitate was dried and then calcined at 500 °C for 3 hours. In such a way, X-ray amorphous samples were obtained. Dispersed crystalline samples were obtained by adding dropwise a solution of nitrates in diluted 1:10 nitric acid to a solution of $NH_4H_2PO_4$. The precipitate was aged at room temperature for 24 h, then centrifuged and washed three times with deionized water, then dried at 90 C and calcined at 500 °C for 3 h. For comparison, a sample with composition $Ca_{0.5}Sr_{0.5}Zr_4P_6O_{24}$~2.06-2.08 sintered at 1300 °C, ball milled and annealed at 1000 °C was used. The water content in samples was estimated using thermal analysis (a Q-1500 device, heating rate 10 degree/min). Bulk structure of all samples was characterized by EXAFS (spectra were acquired at the EXAFS Station of the Siberian Center of Synchrotron Radiation, Novosibirsk using the same procedure as in [5] and fitted with the help of EXCRUV-92 program) and IR spectroscopy of the lattice modes (a BOMEM MB-102 IR-Fourier spectrometer, samples were pressed in wafers with CsI). Crystalline samples were also studied by TEM (JEM 2010 C, 200 keV) and XRD (URD-63 diffractometer, Cu K_a radiation). ^{31}P NMR MAS (a Bruker CXP-300 spectrometer operating at 121,469 Mhz, the MAS rate 2.5 kHz, pulse length 1μs, recycle time 10 or 20s) was used to characterize composition and local environment of the phosphate groups in amorphous and crystalline samples. Surface properties were probed by the IR spectroscopy of surface hydroxyls (Broensted acid centers) and CO test molecule absorbed at coordinately unsaturated cations-Lewis acid centers. In these experiments, samples were pressed in wafers with densities 4.4-22.7 mg/cm and vacuum pretreated in the IR cell at 400 C for 1 h. CO was absorbed at 150-160 K at 5 Torr, and spectra were recorded using a IFS-113 v Bruker spectrometer. Catalytic properties in the reaction of No_x reduction by propane and methane were studied in the plug-flow microreactor in the temperature range 200-500 C at space rates ca 50,000 h and feed composition 300-700 ppm NO_x + 0.17%$CH_4(C_3H_8)$ + 180 ppm SO_2 + 2% H_2O + 5% O_2 + 5% CO_2 in He.

RESULTS

Bulk structure of dispersed low-temperature complex phosphates.

According to thermal analysis data, all samples contained a great amount of water approaching 5-6 molecules for the $MeZr_4P_6O_{24}$ formula unit. Approximately from one third to a half of water is desorbed in a peak at around 120-150 °C, while remaining water is removed continuously at temperatures from 600 to 1000 °C depending on its location within the lattice. Removal of this structural water was followed by exoeffects at 700-800 °C and 900-1000 °C indicating structures relaxation due to annealing of rather highly energetic defects (point defects). In general, structural water can be assigned to different lattice sites/structural groups. Combination of the IR and NMR data allowed one to select the most probable models. Thus, NMR data are in favor of the absence of the acidic phosphate groups in the lattice: only single narrow peak at δ-24.1 (relative to H_3PO_4) was observed for samples sintered at 1300 °C corresponding to PO_4^{3-} groups in a symmetrical environment, while for dispersed samples, superposition of several peaks situated in the range of δ--30-21 and assignable only to phosphate groups in a different environment [6] was revealed. It agrees with the data of the IR spectroscopy of lattice modes which exclude presence of any acidic phosphate groups in samples: a broaden band situated at ~ 1000 cm can be assigned only to v_3 stretching vibration of the PO_4^{3-} group as for the case of a free phosphate ion with the T_d symmetry. Similar spectra of phosphate ions were observed for the ideal heteropolyacids and some layer phosphates [7,8]. Some splitting of this absorption band was observed for crystalline samples indicating variation of the local environment in the lattice. Further, for all samples, absorption bands at 3500 - 3600 cm^{-1} and 1630 cm^{-1} assigned to O-H vibrations in the bulk water molecules were observed. In most cases, strong hydrogen bonds were manifested by a broad absorption in the range from 3000 to 3400 cm^{-1}. These findings indicate that for dispersed samples, water is located in the cationic positions, probably, as H_3O^+ species. Indeed, EXAFS and TEM data support this assumption. For nearly all amorphous and crystalline samples calcined at temperatures up to 700 C, (Fig. 1, Table 1),

202

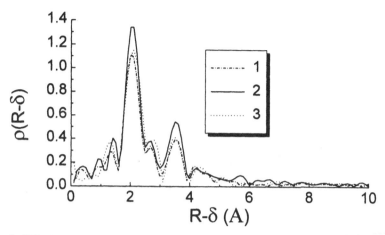

Fig. 1. EXAFS radial distribution curves for hydrated amorphous (1), crystalline (2) $La_{2/3}Zr_4P_6O_{24}$ samples and amorphous $Zr_{0.25}Zr_4P_6O_{24}$ sample (3).

Table 1. EXAFS parameters for some samples

Sample	Interatomic distance R, Å	Coordination number	Peak relation
$La_{2/3}Zr_4P_6O_{24}$, crystalline	2.06	7.9	Zr-O
	3.65	6.0	Zr-P
	3.01	0.4	Zr-La
	3.14	0.4	Zr-La
$La_{2/3}Zr_4P_6O_{24}$, amorphous	2.08	7.5	Zr-O
	3.69	3.6	Zr-P
	3.16	0.4	Zr-La
$La_{0.25}Zr_4P_6O_{24}$, amorphous	2.08	7.3	Zr-O
	3.68	4.3	Zr-P
	2.94	0.5	Zr-Zr

a zirconium cation is surrounded by 6 oxygen anions with Zr-O distance Å, and 1-2 oxygen atoms are situated at a somewhat larger (~0.1 Å) distance. In the ideal structure of the NZP-type, each zirconium cation is surrounded by six oxygen atoms [2]. Hence, for hydrated samples, the increased Zr-O coordination number can be assigned to water molecules retained in the lattice. At a distance ca 3.7 Å, Zr-P peak with a coordination number 6 is present for crystalline sample, that agrees with the ideal structure. For amorphous samples, Zr-P coordination number declines due to disordering. Variation of the distance from Zr to substituting cation (in the case of zirconium phosphate, to another zirconium) from 2.9 to 3.16 Å, as well as variation of the co-ordination numbers suggest distribution of cations between all possible six cation positions available in the parent structure [2] and not usually occupied by sodium in the ideal case.

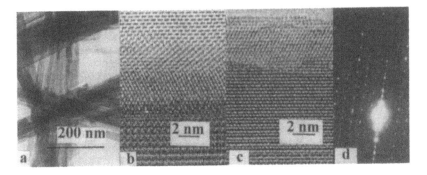

Fig. 2. Typical micrographs of the hydrated crystalline $La_{2/3}Zr_4P_6O_{24}$ sample (a-c) and corresponding microdiffraction (d).

HRTEM observations of dispersed crystalline samples revealed their particles adopted a needle-like shape (Fig. 2a) with typical sizes ~ 100 Å x 1 + 5 μm stacked into bundles with a cross-section ~ 1000 Å. The needles are formed by bands (layers) with a rather perfect ordering of the basic units of NASICON-type structure-two zirconium-oxygen octahedra (Fig. 1b) clearly visible at thin needle edges. This type of local structure agrees well with the EXAFS data. However, the layer stacking across the needle is not perfect, forming a great number of bulk stacking faults and surface steps ~ 10(high (Fig. 2c). Needles are stacked into bundles by side planes with a rather good matching, that is seen from the linear chains of reflexes in SED (Fig. 2d). Stacking faults and interparticle boundaries can generate new types of coordination polyhedra not typical for the ideal structure, which can accommodate water molecules. XRD diffraction from this sample corresponds to a NZP-type structure with a possible presence of some other phases like $NH_4Zr_3P_3O_{12}$.

Surface properties.

Bröensted acid centers. For all samples, in the surface hydroxyls stretching region, the most developed bands are at ~3660 cm^{-1} and are affected by CO adsorption (Table 2). These bands can be assigned to weakly acidic surface P-OH groups. Hence, the surface of all samples is mainly covered by these groups not capable of activating hydrocarbons. This feature most clearly expressed for layered zirconium phosphates. For the most part of samples, more acidic hydroxyls bound with zirconium cations were manifested by less intense bands at ~3745-3765 cm^{-1}. Addition of substituting cations decreases the intensity of the bulk hydroxyls (probably, H_3O^+ species) and increases the relative intensity of Zr-OH groups, probably, via a known charge compensation mechanism (Table 2).

Lewis acid centers. For layered zirconium phosphates, such centers were absent. The same was true for samples containing lanthanum. Addition of the alkaline-earth cations was accompanied by appearance of the surface cus Zr cations (bands at ~2200 cm^{-1}) or Me^{2+} cations (band at ~2170 cm^{-1} for Ca-containing sample). For samples containing copper and iron, new bands typical for CO complexes with Cu^{1+} (v_{co}~2140 cm^{-1}) appeared. For samples containing alkaline-earth cations with a small admixture of copper, the intensity of bands corresponding to Cu^{1+}-CO complexes is too high suggesting copper segregation in the surface layer. For cobalt-containing samples, bands corresponding to $Cu^{+(3+)}$-CO complexes (v_{CO}~2205 cm^{-1}) might be superimposed on those of Zr^{4+}-CO complexes. For samples of the same composition, the density of the surface acid centers varies considerably depending upon the preparation procedure. Thus, amorphous samples usually have a lower density of the acid centers as compared with the crystalline ones. It can be explained by easier relaxation of the surface structure in the former case leading to disappearance of the coordinately unsaturated centers.

Table 2. Main features of the surface acidic centers revealed by IRS.

Samples (cation composition)	Specific surface m²/g	Hydrogen bonded bulk hydroxyls		P-OH		Zr-OH		Carbonyls Me^{n+}-CO	
		a	b	a	b	a	b	a	b
$Zr_{0.25}Zr_4^c$	41	3500	64.5	3655	37	376	2.7	-	-
$CaZr_4^c$	95	3480	22.4	3660	30.7	3765	4.4	2170	9.4
$(CaCu)_{0.5}Zr_4^c$	52	3460	11.2	3660	5.9	3745	1.4	2195	9.3
$La_{2/3}Zr_4^c$	25	3500	40	3655	29.3	3755	3.8	2140	0.8
$FeZr_4^c$	103	-	-	3665	5.3			2183	5.2
$Ca_{0.95}Cu_{0.05}Zr_4^c$	40	3530	42	3660	35.7	3760	3.2	2135	4
$Sr_{0.95}Cu_{0.05}Zr_4^c$		3520	46.4	3655	33.3	-	-	2210	0.5
$Sr_{0.95}Fe_{0.05}Zr_4^c$		3540	39	3655	34	3755	5.6	2140	1
$Sr_{0.95}Co_{0.05}Zr_4^c$		3530	39	3658	30	3750	5.0	2202	1.6

[a]-position of the absorption band (cm⁻¹); [b]-band intensity (absorbance units/g); [c] and [d]-crystalline and amorphous samples, respectively.

Catalytic properties.

Dispersed samples of some complex phosphates were found to be rather active at high space velocities in the reaction of NO_x selective reduction by methane and propane in excess of oxygen and in the presence of such poisons as SO_2 and H_2O (Fig. 3).

The highest activity was exhibited by the dispersed crystalline samples possessing either a high density of Lewis acid sites-cus cation or Broensted acid sites -Zr-OH groups. In our experimental conditions, layered zirconium phosphates do not display any activity, and activity of amorphous samples was either moderate or poor. For the best samples studied here, the level of activity was much higher than that for such typical acid catalyst as H-ZSM-5 [9] while being comparable with activity of such good catalysts as Pd-ZSM-5 or Co-ZSM-5 [9-11]. Due to quite low operational temperatures, NZP-based catalysts can be promising for NO_x removal from diesel exhausts. Though details of the reaction mechanism for dispersed complex zirconium phosphates are still to be clarified, nevertheless, some preliminary ideas can be formulated. Thus, operational temperatures can be explained by the efficient activation of hydro-carbons typical for zirconia-based systems. Further, for these acid systems, such typical intermediates as surface nitrite (NO_2) and nitrile (CN) species [11] are expected to be less strongly bound, and hence more reactive as compared with Co-ZSM-5. Concomitant formation of nitrous oxide in the course of reaction (absent for Co-ZSM-5 and present for Mn-ZSM-5 [11]) suggests a step of interaction between the surface CN species and gas-phase or weakly absorbed NO_2 to be rate-determining.

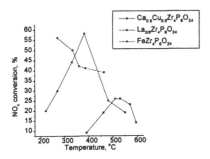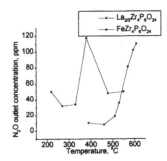

Fig. 3 Temperature dependence of the NO_x conversion (a) and N_2O formation (b) in the course of No_x reduction by methane on the crystalline dispersed complex phosphates. GHSV 50,000 h^{-1} (see experimental part for the other details).

CONCLUSIONS

Bulk and surface microstructures of dispersed complex zirconium phosphates of the NZP type were found to depend upon the sample composition and preparation procedure. Sample acidity characterized by the IR spectroscopy of absorbed CO and surface hydroxyls was shown to broadly depend upon the type of substitutional cations and microstructure. The pronounced role played by the residual water species retained in the cationic positions was emphasized. Some compositions were found to be highly active in the selective catalytic reduction of No_x by hydrocarbons in mixtures simulating real diesel exhausts.

REFERENCES

1. T. Yamaguchi, Appl. Catal. **61**, 1 (1990).
2. J. Alamo and R. Roy, J. Mater. Sci. **21**, 444 (1986).
3. P.A. Agaskar, R.K. Grasselli, D.J. Buttrey and B. White in 3- World Congress on Oxidation Catalysis edited by B.K. Grasselli, S.T. Oyama, A.M. Gaffney and J.E. Lyons, Elsevier Science B.V., 1997, p. 219-226.
4. J.F. Clover, D.K. Agrawal and H.A. McKinstry, J. Mater. Sci. Lett **7**, 422 (1988).
5. D.I. Kochubei, S.N. Pavlova, B.N. Novgorodov, G.N. Kryukova and V.A. Sadykov, J. Catal. **162**, 500 (1996).
6. H. Nakayama, T. Eguchi, N. Nakamura, Sh. Yamaguchi, M. Danjyo and M. Tsuhako, J. Mater. Chem., **7**, 1063 (1997).
7. M.J. Hernandez-Moreno, M.A. Vlibarri, J.L. Rendon and S.J. Serna, Phys. Chem. Mineral. **12**, 34 (1985).
8. C. Rocchiccioli-Deltchiff, R. Thouvenot and R. Frank, Spectrochim. Acta **32A**, 587 (1976).
9. J.N. Armor, Catal. Today **21**, 147 (1995).
10. Y. Nishizaka and M. Misono, Chem. Lett. **1295** (1993).
11. A.W. Aylor, L.J. Lobree, J.A. Reimer and A.T. Bell, J. Catal. **170**, 390 (1997).

STABILITY AND SURFACE ACIDITY OF ALUMINUM OXIDE GRAFTED ON SILICA GEL SURFACE

L.L.L. Prado [*], P.A.P. Nascente [*], S.C. de Castro [**], Y. Gushikem [***]

[*] Centro de Caracterização e Desenvolvimento de Materiais, Departamento de Engenharia de Materiais, Universidade Federal de São Carlos, 1365-905 São Carlos, SP, Brazil
[**] Instituto de Física "Gleb Wataghin", Universidade Estadual de Campinas, 13083-970 Campinas, SP, Brazil
[***] Instituto de Química, Universidade Estadual de Campinas, 13083-970 Campinas, SP, Brazil

ABSTRACT

The synthesis of aluminum oxide grafted on silica gel surface was carried out by the reaction of a suitable aluminum precursor with the surface hydrolysis of the oxide support. The chemical and physical properties of the attached oxide, SiO_2/Al_2O_3, can be quite different than those found for bulk Al_2O_3. The advantage of this preparation method, compared to the conventional ones (impregnation, precipitation and calcination), is that the oxide is highly dispersed on the surface (monolayer or submonolayer). We characterized the surface oxides treated at the temperature range of 423 to 1573 K employing X-ray photoelectron spectroscopy (XPS), solid state nuclear magnetic resonance spectroscopy (NMR), and diffuse reflectance spectroscopy (DRS). XPS was used to identify the oxidation states and atomic ratios. Al^{27} NMR detected two species for samples heated up to 1023 K, and another one above this temperature. DRS, using pyridine as a molecular probe, showed that both Lewis and Brönsted acid sites are stable up to 1023 K. We concluded that the aluminum oxide is highly dispersed on the silica gel surface and it remains stable up to 1023 K.

INTRODUCTION

The study of highly dispersed Al_2O_3 on SiO_2 surface is of interest taking into account its numerous uses, especially in catalytic reactions [1, 2]. Its properties can be quite different than those for bulk phase [3]. The quantity of the exposed active acid centers in monolayer films of metal oxides is considerably higher than for bulk oxide, and the Lewis acid sites are very stable under heat treatments because the oxides are attached to the surface by Si-O-M bonding and are less mobile, avoiding the collapse of the particles and deactivation of the Lewis acid sites [4].

The structures of numerous alumina are known [5-8], but those of highly dispersed oxide structures on a porous substrate solid matrix are less known. In a recent work, a series of high area silica gel substrate treated with well characterized aluminum solutions was prepared, and the structure of the obtained SiO_2/Al_2O_3 was studied by Magic Angle Spinning (MAS) ^{27}Al NMR spectroscopy [9].

In this work, the preparation of SiO_2/Al_2O_3 was carried out by reacting SiO_2 with $Al(OR)_3$ [R= $(CH_3)_2CH$-] in non aqueous solvent. The reaction of alkoxide with silanol groups, $\equiv SiOH$, occured readily, yielding Al-O-Si bond formation. In this case, only monolayer or sub-monolayer could be formed on the substrate surface. The structure, thermal stability and acidity of the grafted species are discussed.

EXPERIMENT

Aluminum oxide was grafted on silica gel surface by immersing 70 g of silica gel (Aldrich, $S_{BET}= 500$ m^2.g^{-1}, average diameter pore of 6 nm, and particle sizes between 0.2 an 0.05 mm), previously heated at 423 K under vacuum (10^{-3} torr), in 250 ml of 0.14 M toluene solution of aluminum isopropoxide (Aldrich). The suspension was refluxed under nitrogen atmosphere for 24 h. The mixture was filtered under nitrogen atmosphere, washed with dry ethanol and heated under vacuum in order to eliminated all solvent. Furthermore, the solid was immersed in bidistilled water to promoted the hydrolysis of the unreacted Al-O-R bond, filtered, washed with water, and dried at 393 K under vacuum for 8 h. The following chemical equations describe the preparation reactions:

$$n\equiv SiOH + [(CH_3)_2CHO]_3Al \rightarrow (\equiv SiO)_nAl[OCH(CH_3)_2]_{3-n} + n(CH_3)_2CHOH \qquad (1)$$

$$(\equiv SiO)_nAl[OCH(CH_3)_2]_{3-n} + (3-n)H_2O \rightarrow (\equiv SiO)_nAl(OH)_{3-n} + (3-n)(CH_3)_2CHOH \qquad (2)$$

where $\equiv SiOH$ stands for silanol group on silica gel surface.

The quantity of Al incorporated in $(\equiv SiO)_nAl(OH)_{3-n}$, hereafter designated as SiO_2/Al_2O_3, was determined by X-ray fluorescence spectroscopy. The quantity of aluminum grafted on the silica gel surface was 1.92 wt%, corresponding to 0.73 mmol.g^{-1}.

Fine-grained samples of SiO_2/Al_2O_3 were heat treated for 8 h at the following temperatures: 423, 573, 873, 1173, 1323 and 1473 K. The specific surface areas of the samples were obtained by the BET multipoint method on a ASAP 2010 equipment.

The XPS analyses were carried out on a Kratos XSAM HS spectrometer, under 10^{-9} torr, using Mg Kα radiation as X-ray source. The spectra were referenced to the binding energy of 284.8 eV for the C 1s line of adventitious carbon.

The ^{27}Al MAS NMR spectra were obtained using a Bruker AC 300/P spectrometer. The chemical shifts of the samples were calibrated against a external AlCl$_3$ solution.

The infrared spectra were obtained on a Bomen MB series using a Jasco DR 81 diffuse reflectance accessory. About 0.5 g of SiO_2/Al_2O_3 was previously degassed at 10^{-3} torr and furthermore submitted to a pyridine vapor for few minutes, and the excess eliminated under vacuum for 4 h at 473 K.

RESULTS AND DISCUSSION

The specific surface area of the untreated silica is 460 m^2.g^{-1}, and that of SiO_2/Al_2O_3 not submitted to heat treatment is 416 m^2.g^{-1}. There is an area decrease upon chemical modification of the surface, presumably due to the smallest pores being blocked by the reagent. Figure 1 shows the variation of the specific areas as the samples were heat treated from room temperature up to 1473 K. A large area decrease is observed for sample treated at 1023 K, and above 1100 K the matrix becomes non-porous.

Table I summarizes the XPS results. Two oxygen O 1s peaks can be distinguished: one between 531.1 and 532.3 eV due to oxygen bonded to silicon, Si-O (designated as O^I) [10], and other between 532.6 and 534.2 eV due to oxygen bonded to aluminum, Al-O (designated as O^{II}) [11]. The numbers in parenthesis are the atomic percentage of each contribution to the O 1s peak. As the calcination temperature increases, the atomic percentage of O^I increases while the O^{II} decreases, and above 1023 K only O^I is detected. The Al/Si ratios decrease as the calcination temperatures increases, presumably due to a diffusion of the aluminum metal to the interior of the matrix . The Si 2p binding energies are observed between 102.8 and 103.3 eV, and Al 2p binding energies, between 74.6 and 75.2 eV. There is not a detectable phase separation due to sample calcination at higher temperatures.

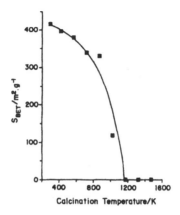

Figure 1. Variation of the specific area with the calcination temperature.

Table I. XPS data for SiO$_2$/Al$_2$O$_3$ heat treated at various temperatures.

T (K)	Binding Energies (eV)			Atomic Ratios		
	O 1s	Si 2p	Al 2p	Al/Si	Al/O	Si/O
273	531.1 (32%) 532.6 (68%)	103.1	74.7	4.55	0.30	0.067
423	531.5 (51%) 532.8 (49%)	102.8	74.6	2.51	0.35	0.14
573	531.9 (86%) 533.6 (14%)	103.0	74.9	2.29	0.31	0.14
723	532.0 (86%) 533.4 (14%)	103.1	75.0	2.34	0.34	0.14
873	531.9 (87%) 533.6 (13%)	103.2	75.1	2.87	0.37	0.13
1023	532.3 (95%) 534.2 (5%)	103.3	75.2	2.73	0.36	0.13
1173	532.3	103.3	75.2	2.61	0.41	0.16
1323	532.3	103.2	75.1	1.22	0.27	0.22
1473	532.3	103.2	75.0	1.53	0.30	0.19

Figure 2 shows the NMR spectra for the SiO$_2$/Al$_2$O$_3$ sample as obtained and dried under vacuum and for samples calcined between 423 and 1573 K. The peak at 10 ppm is due to the aluminum species in an octahedral environment, and the peak at 55 ppm is due to aluminum in a tetrahedral environment, both corresponding to a Al$_{oct}$-(OSi)$_4$ and Al$_{tetr}$-(OSi)$_4$ environments, respectively [9]. For samples calcined between 1173 and 1573 K, an additional peak at 35 ppm is observed. For bulk Al$_2$O$_3$ this band is assigned to a pentacoordinated aluminum, more specifically for ground bohemite calcined at 973 K [12]. However, in the SiO$_2$/Al$_2$O$_3$ sample prepared by mixing SiO₂ and soluble ionic aluminum in aqueous solution, this peak was not observed for either calcined or uncalcined samples [9]. In the present case, the appearance of this peak can be associated to phase separated aluminum oxide on the surface, which could not be detected by XPS.

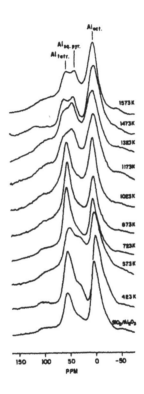

Figure 2. NMR spectra for the SiO₂/Al₂O₃ sample as obtained and dried under vacuum and for samples calcined between 423 and 1573 K.

The Lewis acid sites, as well as their stability, were studied by sorbing pyridine as a probe molecule on the precalcined samples. The recorded spectra are shown in Figure 3. The weak band observed at 1600 cm⁻¹ and other one at 1450 cm⁻¹ are assigned to the 8a and 19b vibrational modes of pyridine coordinated to a Lewis acid site [4, 13]. For samples precalcined at temperatures between 573 and 1023 K, the Lewis acid sites are observable, and for higher temperatures, the specific surface area was so drastically reduced (near 1 m².g⁻¹) that pyridine molecule was not detected.

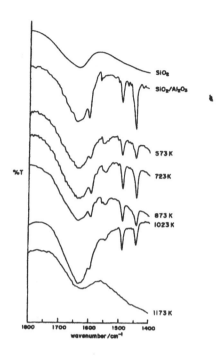

Figure 3. Diffuse reflectance infrared spectra using pyridine as a molecular probe.

CONCLUSIONS

We have investigated the structure, thermal stability and acidity of aluminum oxide grafted on silica gel surface. We characterized by BET, XPS, NMR and DRS the fine-grained samples of SiO_2/Al_2O_3 heat treated at the temperature range of 423 to 1573 K. The specific surface areas decreased with the increase in the calcination temperature, manifesting an abrupt decrease for the sample treated at 1023 K. Above 1100 K the matrix became non-porous. DRS, using pyridine as a molecular probe, showed that the Lewis acid sites were observable for samples calcined at temperatures up to 1023 K, and for higher temperatures, the specific surface area was so drastically reduced that pyridine was not detected. The Al/Si atomic ratios measured by XPS remained approximately constant for samples heat treated at the temperature range of 423 to 1173 K, but decreased significantly for higher calcination temperatures, presumably due to the diffusion of aluminum to the matrix interior. NMR detected two species for samples calcined between 423 and 1023 K, and a third one for higher calcination temperatures, which could be associated to phase separated aluminum oxide on the surface. We concluded that the aluminum oxide was highly dispersed on the silica gel surface and it remained stable up to 1023 K.

ACKNOWLEDGMENTS

This work was supported by the Brazilian agencies FAPESP and CNPq.

REFERENCES

1. R.S. Drago and E.E. Getty, J. Am. Chem. Soc. **110**, p. 3,311(1988).

2. R.S. Drago, S.C. Petrosius and C.W. Chronister, Inorg. Chem. **33**, p. 367 (1994).

3. D.W. Goodman, J. Vac. Sci. Technol. **A14**, p. 1,526 (1996).

4. S. Denofre, Y. Gushikem, S.C. de Castro and Y. Kawano, J. Chem. Soc. Faraday Trans. **89**, p. 1,057 (1993).

5. S.N.R. Rao, E. Waddel, M.B. Mitchell and M.G. White, J. Catal. **163**, p. 176 (1996).

6. E.C. De Canio, J.C. Edwards and J.W. Bruno, J. Catal. **148**, p. 176 (1994).

7. D.L. Cocke, E.D. Johnson and R.P. Merrill, Cat. Rev. Sci. Eng. **26**, p. 163 (1984).

8. H. Knozinger and P. Ratnasamy, Cat. Rev. Sci. Eng. **17**, 31 (1978).

9. W.E.E. Stone, G.M.S. El Shafei, J. Sanz and S.A. Selim, J. Phys.Chem. **97**, p. 10,127 (1993).

10. A. Hess, E. Kemnitz, A. Lippitz, W.E.S. Unger and D.H. Menz, J. Catal. **148**, p. 270 (1994).

11. C.D. Wagner, D.E. Passoja, H.F. Hillery, T.G. Kiniski, H.A. Six, W.T. Jansen and J.A. Taylor, J. Vac. Sci. Technol. **21**, p. 933 (1982).

12. F.R. Chen, J.G. Davis and J.J. Fripiat, J. Cat. **133**, p. 263 (1992).

13. C. Morterra and G. Magnacca, Cat. Today, **27**, p. 497 (1996).

Part VII

Surface Modifications

DESIGN OF CATALYSTS WITH ARTIFICIALLY CONTROLLABLE FUNCTIONS USING SURFACE ACOUSTIC WAVES AND RESONANCE OSCILLATIONS GENERATED ON FERROELECTRIC MATERIALS

N.Saito, Y.Ohkawara, K.Sato, and Y.Inoue*
Department of Chemistry, Nagaoka University of Technology, Nagaoka 940-2188
Japan

ABSTRACT

A poled ferroelectric material of a $LiNbO_3$ single crystal and a polycrystalline lead strontium zirconium titanate(PSZT) disc were employed as a catalyst substrate, on which various metals such as Pd, Ag, Ni and Pt and a WO_3 oxide were deposited as a thin film catalyst. The effects of acoustic waves generated by rf power upon catalyst activation and reaction selectivity have been examined. The activity for CO oxidation of a Pt/PSZT catalyst increased by a factor of 3.9 with a thickness-extensional(TE) mode of resonance oscillation(RO) and 2.9 with a radial-extensional (RE) mode. The activation energy of the reaction was larger for the TE mode(38 $kJmol^{-1}$) than that for the RE mode(24 $kJmol^{-1}$). The polarization axis-dependent changes in surface potential occurred with the TE mode, but not with the RE mode. With the TE mode, the activation energy of ethanol oxidation over a $Ni/LiNbO_3$ and a $Ag/LiNbO_3$ catalyst decreased from 36 to 19 $kJmol^{-1}$ and from 76 to 30 $kJmol^{-1}$, respectively, and the TE mode had a larger effect on the reaction with a higher activation energy. In ethanol dehydrogenation and dehydration on a WO_3 / $LiNbO_3$ catalyst, the TE mode increased remarkably the selectivity for ethylene production without giving a significant change to acetaldehyde production. The effects of acoustic wave excitations on the catalysts are discussed.

INTRODUCTION

The design of a catalyst surface which possesses artificially controllable functions for chemical reactions has been among interesting subjects in heterogeneous catalysis, but only a few studies have so far been performed. Since catalysis by metal or oxide surfaces is closely associated with the arrangement of surface atoms and the surface electron density, it is useful to change both factors in an artificial manner. For this purpose, it is an interesting appoach to use dynamic lattice displacement induced on catalyst surfaces. We have paid attention to a piezoelectric phenomenon of a ferroelectric crystal with spontaneous polarization[1].

Surface acoustic waves(SAWs) generated on poled ferroelectric crystals by applying radio frequency power cause time-dependent lattice displacement, and it has been demonstrated that the SAWs have significant effects on the enhancement of catalytic activities[2-9].

An another interesting method is resonance oscillations(ROs) which can be generated at a defined frequency and have different vibration modes such as a radial extensional(RE) and a thickness extensional(TE) mode. Recently, using a poled polycrystalline ferroelectric $Pb_{0.95}Sr_{0.05}Zr_{0.53}Ti_{0.47}O_3$ (PSZT) substrate, we have shown that the RE and TE mode of RO have different effects on catalytic activity for ethanol oxidation of a Pd thin film catalyst deposited on the substrate[10-14]. More recently, with a single crystal of z-cut lithium niobate, the TE mode of RO gives rise to anomalous enhancement of the catalytic activity for the same reaction on Pd[15]. From a viewpoint of approach to new catalyst materials, it is of importance to investigate the effects of the RE and TE modes on different kinds of the catalytic reactions over various catalysts. This is also useful for a better understanding of a surface acosutic wave(SAW)- and a RO-induced catalyst activation mechanism. The present paper deals with the RO effects on ethanol oxidation over Ni and Ag catalysts and on CO oxidation on Pt in order to reveal the different effects of the TE and RE mode. The results are discussed in comparsion with the SAW effects. Changes in surface properties with the RE and TE mode were investigated by measuring surface potential. The same measurement was also applied to a surface on which Rayleigh SAW propagates.

The previous study on ethanol oxidation over Pd and Ni catalysts has shown that the SAW effects are larger when catalysts surfaces are covered with strongly adsorbed oxygen or oxide layers[3,4]. In this regard, the effects of RO on metal oxide catalysts are interesting, and WO_3 was chosen as a catalyst in the present work. In order to obtain information about RO effects on selectivitry, dehydrogenation and dehydration of ethanol was investigated.

EXPERIMENT

A poled ferroelectric polycrystalline $Pb_{0.95}Sr_{0.05}Zr_{0.53}Ti_{0.47}O_3$ (referred to here as PSZT) and a poled ferroelectric single crystal of z-cut $LiNbO_3$(referred to as z-LN) were employed as a substrate. These crystals have a polarization axis normal to surface, thus exposing a positively polarized surface at one plane and a negatively polarized surface at the other. PSZT was in the form of a disc(25 mm in diameter and 0.22 mm in thickness) and showed resonance line of the RE mode at a frequency of 80(first), 209(second), 327(third) kHz and the TE mode at 10, 30 and 50 MHz. A substrate of z-LN has a rectangle with 1.0 mm in thickness, and its TE mode appeared at a frequency of 3.4(First), 10.5(second) and 17.4(third) MHz. In the present work, the first resonance frequencies were used for catalyst activation, unless otherwise specified. Both planes of a PSZT disk were covered with Ag paste electrodes for input of high frequency electric power. The Ag paste electrodes were inactive for the catalytic reactions under the present reaction conditions. A Pt film was deposited on PSZT by an evaporation method with electron beam heating of a pure Pt metal in high vacuum(denoted here as Pt/PSZT). For z-LN, the front and back planes were covered with a catalytically active Ni and Ag thin film which functions as both electrodes and catalysts. The films were deposited at a thickness of 100 nm by resistance-heating of the respective metals in vacuum(referred here to as Ni/z-LN and Ag/z-LN, respectively).

High frequency electric power was generated from a network analyzer(Advantest, R3751AS), amplified(Kalmus, AS225LC), and then introduced to a catalyst sample. Catalyst temperature was monitored by a radiation thermometer(JSC,TSS-5N) through a BaF2 window and controlled by an outer electric furnace. The catalytic reactions such as ethanol and CO oxidation, and dehydrogenation and dehydration of ethanol were carried out in a gas-circulating apparatus, and the products were analyzed by a gas chromatograph connected to the reaction system.

The contact potential difference of the catalyst surface was measured in air at room temperature by a dynamic condenser electrometer(Ando Elec. Co., AA2404). A sample was placed in parallel with a probe of the electrometer at a distance of 3 mm. A voltage generated on the Ag surface was monitored in the absence and presence of the TE and RE mode.

The structure of a 20 MHz SAW device was described elsewhere[2-4]. Briefly, a poled ferroelectric single crystal employed was 128°-rotated Y cut $LiNbO_3$ for the propagation of Rayleigh SAW. For the input and output of radio frequency(rf) power, the interdigital transducer(IDT) electrodes of Al were photolithographically fabricated on both ends of the ferroelectric crystal surface. For measurements of surface potential, a Ni catalyst thin film was deposited in the middle of the two IDTs by resistance heating.

RESULTS

RO effects on CO oxidation and changes in surface potential

Figure 1 shows the CO oxidation on a Pt/PSZT catalyst. With RO-off, CO_2 was produced at a constant rate. Upon the introduction of the RE mode at a resonance frequency power of 3 W at 80 kHz, CO_2 production increased immediately, and higher activity was maintained until the power was turned off. The production of CO_2 returned to the original level with RO-off. Similar response in CO_2 production was observed for the introduction of the TE mode at a frequency of 10 MHz, but an increase in the CO_2 production was larger, compared to that by the RE mode. Activation coefficient, V_{on}/V_{off}, defined as the ratio of activity with RO-on to that with RO-off is 3.9 for the TE mode and 2.9 for the RE mode.

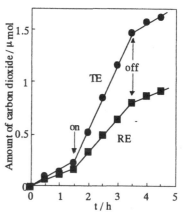

Fig.1 Changes in CO_2 production on a Pt/PSZT catalyst with RE and TE mode
P=3W,T=373K

Figure 2 shows the temperature dependence of the reaction rate with the RE and TE mode. The activity over this temperature range was higher for the TE mode. The temperature dependence was different between the two modes:the activation energy was 38 kJ mol^{-1} for the TE mode, whereas it dropped to 24 kJmol^{-1} for the RE mode.

Figure 3 shows the influences of the RE and TE mode on the surface potential. With the RE mode, no significant change in the surface potential was observed. On the other hand, the TE mode produced a positive voltage for the negative plane and a negative voltage for the positive plane. With the TE mode-off, the induced voltage disappeared. Any frequency difference from resonance frequency caused no changes in the surface potential. The similar measurements were performed for a Ni deposited SAW surface, and no change in the surface potential was observed during Rayleigh SAW propagation.

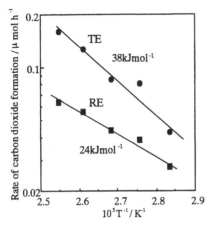

Fig.2 Comparison of temperature dependence
of CO oxidation between RE and TE mode
P=3W

Fig.3 Changes in surface potential with RE and
TE mode for (+) and (-)PSZT plane
P=3W,T=room temperature

RO effects on ethanol oxidation

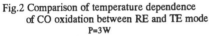

A ferroelectric single crystal of z-LN was employed as a substrate for RO generation. Figure 4 shows the temperature dependence of ethanol oxidation on Ni and Ag thin film catalysts. For a Ni/z-LN, the TE mode caused an increase in the catalytic activity by a factor of 6 at 423K and a decrease in activation energy from 36 to 19 kJ mol⁻¹. A larger activity increase was observed for a Ag/z-LN catalyst: the value of V_{on}/V_{off} was 14 at 373K. With the TE mode, the activation energy decreased from 76 to 30 kJmol⁻¹. Table 1 summarizes the effects of the TE mode on catalyst activation and activation energy of ethanol oxidation for the three metal catalysts by adding the previous results obtained for Pd/z-LN[13].

Figure 5 shows changes in V_{on}/V_{off} with rf power. For a Ag/z-LN catalyst, V_{on}/V_{off}

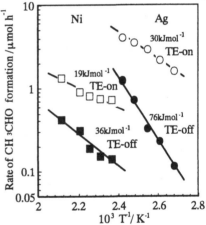

Fig.4 Temperature dependence of ethanol
oxidation on Ag/z-LN and Ni/z-LN
catalysts.
RO:TE mode,P=3W,fr=3.4MHz

was small below 1 W, increased gradually with increasing power, and steeply above 2 W. The value of V_{on}/V_{off} was nearly proportional to the power of 1.8 of applied electric power. For a Ni/z-LN catalyst, a similar increase in V_{on}/V_{off} occurred as a function of rf power, but the increase in V_{on}/V_{off} above 2 W was moderate.

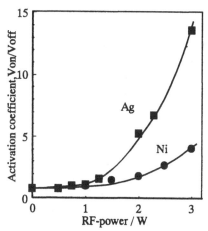

Fig.5 Activation coefficient,Von/Voff, for
ethanol oxidation as a function of power.
T_{Ag}=373K,T_{Ni}=443K,fr=3.4MHz

Dehydrogenation and dehydration of ethanol

In the dehydrogenation and dehydration of
ethanol on WO_3/z-LN at 553 K, both
ethylene and acetaldehyde were formed
from an initial stage. With the TE mode-on,
the production of acetaldehyde increased
slightly, whereas that of ethylene increased
to a considerable extent. With the TE
mode-off, the formation of these products
returned to their original levels. Figure 6
shows the effects of power on the activity
coefficients for the production of ethylene
and acetaldehyde. The coefficient for the
acetaldehyde production increased to 2 at a
power of 0.5 W and remained constant
with increasing power. On the other hand,
the coefficient for ethylene production
increased linearly with power. At a power
of 3 W, the coefficient for ethylene
production was 6 fold larger than that of
acetaldehyde production.

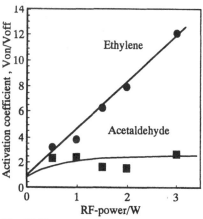

Fig.6 Difference in Von/Voff between ethylene
and acetaldehyde productions with power.
T=553K,fr=3.4MHz

DISCUSSION

An interesting feature of CO oxidation on a Pt/PSZT catalyst is that there is a considerable difference not only in catalyst activation, but also in activation energy of the reaction between the RE and TE modes. This catalytic difference is attributed to the fact that the TE mode is able to cause changes in the surface potential, whereas the RE mode produces little change. With respect to the effects of polarized surfaces, it is to be noted that a positive voltage is generated on the (-) plane, and a negative voltage on the (+) plane with the TE mode, whereas no significant changes take place with the RE mode. A recent study on ethanol oxidation over Ag deposited on either (+)PSZT or (-)PSZT has demonstrated that the activation energy of the reaction between them was different with the TE mode, but there was no significant difference with the RE mode[14]. Furthermore, for ethanol oxidation over a Pd/z-LN catalyst, the TE mode gave rise to a higher activation energy of the reaction over Pd on (+)z-LN than on (-)z-LN[15]. It is evident that irrespective of kinds of the catalytic reactions and of the catalysts, there is a clear difference in catalyst activation between the TE and RE mode. Such a difference is considered to be due to the fact that the TE mode involves changes in the electronic structures of surface atoms, which occurs, to a lesser extent, with the RE mode. Namely, the appearance of changes in the surface potential suggests that the TE mode has influence on the electronic states of surface metal atoms.

In the CO oxidation on a metallic Pd surface, Rayleigh SAW caused a decrease in the reaction order from 0.4 to 0.2 with respect to CO pressure, whereas it gave little change in the reaction order with respect to oxygen pressure(the order was 0.8)[7]. These results indicate that CO is more strongly adsorbed than oxygen, and the SAW produces slightly stronger adsorption of CO. Since no changes in the surface potential occurred during Rayleigh SAW propagation, it is likely that the RE mode leads to the stronger CO adsorption as observed in the SAW, and it is expected that the TE mode gives rise to larger changes in kinetic parameters.

For ethanol oxidation, as shown in Table 1, the activation energy attenuated with the TE mode, and the extent of its decrease was 92% for Pd, 61% for Ag and 47% for Ni. The value of V_{on}/V_{off} increased in the order of Pd>Ag>Ni at 373 K. These results indicate that the TE mode has larger effects on catalysts which provide higher activation energy. A laser Doppler method was used to measure vertical lattice displacement during SAW propagation, and it has been demonstrated that the displacement increases in proportion to square root of electric power, which is in good agreement with Auld's equation[16]. Relations between V_{on}/V_{off} and rf power, shown in Fig. 5, demonstrate that the activity enhancement is proportional to the power of 1.6 for the Ni/z-LN and 1.8 for the Ag/z-LN in the higher range of rf power. If one assumes that displacement occurs according to Auld's equation, these results indicate that the activity enhancement occurs with a large nonlinear feature(the power of 3.6 - 3.8) with respect to lattice displacement. This would be acceptable if one considers changes in the chemical bond broadening .

Table1 Changes in activation energy of reaction with TE mode

metal	substrate	Ea / kJmol^{-1} without TE	with TE	Extent of decrease in Ea / %
Pd		156	12	92
Ag	z-LN	76	30	61
Ni		36	19	47

Ea : activation energy

The previous studies on the SAW effects upon ethanol oxidation over Pd and Ni catalysts have shown that oxygen is strongly adsorbed on the catalyst surfaces[3,4]. On the basis of the kinetic reaction orders, a rate-determining step is proposed to be a surface reaction between a dissociatively adsorbed oxygen atom and an adsorbed ethanol to produce an adsorbed acetaldehyde and water:

$$O(a) + C_2H_5OH(a) \rightarrow CH_3CHO(a) + H_2O(a)$$

where the symbol (a) represents an adsorbed state. The SAW is proposed to have a significant influence on the strongly adsorbed oxygens rather than the weakly adsorbed ethanol. Furthermore, in a recent study on ethanol oxidation over a Pd/(+) z-LN surface, the TE mode was found to produce stronger adsorption of oxygen and weaker adsorption of ethanol[15]. In a thermal desorption study on single crystals, Madix et al showed that the state of oxygen is important to the removal of a hydrogen from adsorbed ethanol[17]:the negatively charged surface oxygens promote the process. The changes in the kinetic behavior of ethanol with SAW-on are explained by assuming that the SAW causes a larger change in strongly adsorbed than weakly adsorbed reactants and produces negatively charged surface oxygens so as to promote the abstraction of the hydrogen atom from the adsorbed ethanol. It is likely that a similar situation is invoked for the TE mode.

The effects of the TE mode on the oppositely polarized PSZT surfaces are considered to be due to the combined effects of sonic wave-electron interactions and spontaneous polarization of PSZT. For the TE mode, the direction of vibration is perpendicular to the PSZT surface, and the interactions with the sonic wave excite electrons or promote their movement. Because of a strong field due to the polarization axis, the electrons are accumulated on the positive PSZT plane. The film catalyst in contact with these planes is affected so that the catalyst on the positive PSZT results in the formation of a negatively charged state. The opposite change occurs with a negative PSZT plane, thus giving rise to an electron-deficient state. On the other hand, the direction of vibration in the RE mode is parallel to the surface, for which the movement of the electrons toward surface is not induced. This leads to little change in the surface potential with the RE mode. Furthermore, an additional effect is a difference in resonance frequency between the RE and TE mode: the former(80 kHz) is ca.100 fold lower than the latter(10 MHz), which gives rise to weaker interactions of the RE mode with the electrons. A recent study using low energy photoelectron spectroscopy has shown that a threshold energy for photoelectric emission from a Ag film deposited on PSZT varies considerably with the TE mode . For Ag on the positive PSZT, a threshold energy shifted to lower energy side by 0.12 eV with the TE mode, thus indicating a decrease in work function of the Ag surface. Note that this change occurs under the conditions of the negatively charged surfaces[14]. These results indicate that the TE mode influences the density of electrons at catalyst surface, which is considered to be responsible for catalyst activation. It is likely that higher electron density is associated with the negatively charged oxygens to such an extent that the removal of a hydrogen atom from an adsorbed ethanol is facilitated.

On the other hand, the RE mode causes little changes in surface potential, but catalyst activation occurs. A plausible explanation is that the arrangement of surface atoms is induced by lattice displacement due to the RE mode. However, the total lattice displacement of a PSZT disc is estimated as a few microns in the RE mode, and it is unlikely that bond distance between metal surface atoms is increased significantly by the lattice displacement, provided that lattice displacement occurs uniformly over the crystal plane. The Ag thin film employed is polycrystalline whose surfaces are composed of many imperfections such as grain boundaries, dislocations, and vacancies. The acoustic wave is likely to be concentrated on the imperfect sites, which results in changes in the arrangement of the specific local structures. The characteristic difference between the TE and RE mode is explained in terms of a view that the TE mode induces the larger contributions of electronic structures, whereas the RE mode favors the geometric contributions.

An increase in the selectivity of ethylene production with the TE mode on the WO_3 catalyst indicates that the abstraction of OH group is accelerated, compared to that of hydrogen molecules. This suggests that strong interactions of WO_3 surface with oxygen atoms are induced with the TE mode. As described above, in ethanol oxidation, the TE mode brings about the strong adsorption of oxygen, which gives support to the presence of strong interactions of oxygen-involving group OH with catalyst surfaces.

In conclusion, the effects of RO and SAW on catalysts are based on the same principles of dynamic lattice displacement and on resulting changes in the surface potential of the catalyst surfaces. These methods are proposed to show promise for the development of heterogeneous catalysts with artificially controllable functions.

ACKNOWLEDGEMENT

This work was supported by a Grant-in-Aid for Scientific Research on Priority Areas and for Exploratory Research from The Ministry of Education, Science, Sports and Culture.

REFERENCES

(1) T. Ikeda, in Fundamentals of Piezoelectricity, Oxford Univ. Press. Oxford , pp5-28(1990)
(2) Y. Inoue, M. Matsukawa, and K. Sato,   J. Am. Chem. Soc., 111, 8965 (1989)

(3) Y. Inoue, M. Matsukawa, and K. Sato, J. Phys. Chem., 96 , 2222 (1992)

(4) Y. Inoue, Y. Watanabe, and T. Noguchi, J. Phys. Chem., 99, 9898(1995)

(5) M.Gruyters, T.Mitrelias, and D.A.King, Appl. Phys., A61,243(1995)

(6) Y.Watanabe, Y.Inoue, and K. Sato, Chem. Phys. Lett., 244, 231(1995)

(7) Y. Watanabe, Y. Inoue, and K. Sato, Surf. Sci., 357/358, 769(1996)

(8) H. Nishiyama, M. Shima, N. Saito, Y. Watanabe, and Y. Inoue, Faraday
    Discussion 107, in press.

(9) S. Kelling, T. Mitrelias, J. Gu, V.P. Ostanin, and D.A. King, Faraday
    Discussion 107, in press.

(10) Y. Inoue, J. Chem. Soc., Faraday Transactions, 90 , 815(1994)

(11) Y. Inoue, and Y. Ohkawara, J. Chem. Soc. Chem. Commun., 2101(1995)

(12) Y.Ohkawara, N. Saito, and Y.Inoue,   Sur. Sci.,   357/358 , 777 (1966)

(13) N.Saito,Y.Ohkawara,Y.Watanabe,and Y.Inoue, Appl. Surf. Sci.,121/122,343(1997)

(14) Y.Ohkawara, N.Saito, K.Sato, and Y.Inoue, Chem. Phys. Lett., in press.

(15) N.Saito, Y.Ohkawara, and Y.Inoue, submitted to Surf. Sci..

(16) B.A.Auld, in Acoustic Fields and Waves in Soilds   vol II, Wiley Interscience Pub.,
    John Wiley & Sons Inc., New York , pp.278(1973)

(17) M.A.Barteau, and R.J.Madix, Surf.Sci.,120,262(1982)

HUMIDITY EFFECTS ON THE ELECTRICAL PROPERTIES
OF EPITAXIAL RUTILE THIN FILMS

D.R. Burgess[1], P.A. Morris Hotsenpiller[2], O. Kryliouk[1], and T.J. Anderson[1]
[1]University of Florida, Dept. of Chemical Eng., Gainesville, FL 32611-6005
[2]DuPont Co., Experimental Station, Wilmington, DE 19880-0356

ABSTRACT

Thin films of (001) and (100) oriented rutile phase TiO_2, undoped or doped with Ga or Nb, have been grown using the MOCVD technique on sapphire substrates for use in studies of the effects of humidity on the electrical properties of rutile. The crystallographic and microstructural quality of the films decreases with increasing Ga and Nb concentrations. Heteroepitaxy is, however, maintained with Ga or Nb concentrations up to 4.5 at% for the (001) orientation and to 0.5 at% Ga for the (100) orientation. The electrical properties of the (001) oriented rutile films have been characterized from room temperature to 225 °C in dry and humid, N_2 and air atmospheres. At constant temperature in dry atmospheres, the conductance of the Nb-doped rutile films is greater than that of the undoped, which is greater than the conductance of the Ga-doped films. The activation energies for conduction in the Nb-doped and undoped rutile films in dry atmospheres are similar (~0.1 eV), whereas the activation energy in Ga-doped films is much greater (~0.8 eV). The effects of humidity in reducing the resistance of rutile is greatest in the Ga-doped and very thin (~150 Å) undoped films. Humidity is observed to have similar effects on both the (001) and (100) oriented 0.5 at% Ga-doped films.

INTRODUCTION

Rutile phase TiO_2 thin films have a variety of practical applications. In addition to their use as waveguides, photocatalysts and components in electrical devices, they are also of interest as chemical sensors. Typically, chemical sensors are porous polycrystalline ceramics to maximize the surface area which can interact with the gas phase [1-3]. Additives are used to change the microstructural characteristics and increase the sensitivity of the material to the gas being detected [4-7]. In many sensor structures, however, the exact nature of these changes is not well known. We are using oriented rutile thin films doped with acceptor (Ga) and donor (Nb) ions to examine the effects of surface orientation, space charge potential and surface morphology on the chemical sensing characteristics of rutile. This paper reports the results of our study on the effects of humidity on the electrical properties of undoped and doped (Ga, Nb), (001) and (100) oriented rutile thin films grown by the MOCVD technique on sapphire substrates.

EXPERIMENTAL

The films were grown by MOCVD in a high vacuum, inverted pedestal, hot-walled reactor. The growth temperature was 650°C and the growth rate was nominally 40 Å/min resulting in a typical film thickness of 3500 Å. Solid precursors (2,2,6,6-tetramethyl-3,5-heptanedionato) were used as sources of Ti, Ga, and Nb. Doping was accomplished by mixing dopant precursor material with the Ti precursor on a per Ti basis. For further details regarding growth, please see reference [8]. The crystallographic, microstructural and surface characteristics of the films were examined using X-ray diffraction and atomic force microscopy (AFM). The doping concentrations and distributions in the films were determined using Rutherford backscattering spectrometry (RBS) of 2 MeV $^4He^+$ ions and secondary ion mass spectrometry (SIMS).

The electrical measurements were carried out in an environmental chamber in dry and humid, N_2 and air atmospheres. Measurements were made at 1 atmosphere in flowing gas at temperatures ranging from 25 to 225 °C. High-purity (99.9995%) N_2 and air were used for the dry atmospheres. Humid atmospheres were obtained by bubbling the gases through water at room temperature. The gas flow rate was 150 sccm resulting in the stream containing ~3 mol% H_2O, assuming saturation. The chamber was evacuated to 10^{-6} Torr to purge the atmosphere before changing to a dry or different gas.

The electrical properties were determined using two-probe AC conductivity measurements over a 1 Hz to 10 MHz frequency range. Gold strip electrodes (8mm x 0.5 mm x 2 mm apart) were evaporated onto the surfaces of the films. Conductivity measurements were repeated under each environmental condition and temperature until a constant value was obtained.

RESULTS AND DISCUSSION

To understand the electrical properties and the effects of humidity on the rutile thin films, the films' microstructures must be well understood. Table I summarizes a number of details concerning the crystallography and microstructure of the doped and undoped films used in this study. The left-most column shows the two rutile film orientations, (001) and (100), that are the focus of the current study with their respective sapphire substrates. Listed below each film/substrate pairing are the matching in-plane film/substrate directions. The percentage listed to the right of each in-plane direction pair is the lattice mismatch in that direction for an undoped film at room temperature. The second column lists the acceptor (Ga) and donor (Nb) dopant concentrations investigated. The next two columns list the full width at half maximum (FWHM) of Θ and Φ X-ray diffraction peaks, indicating the alignment perpendicular and parallel to the plane of the film. For the (001) orientation, the Θ scan is of the (002) rutile peak ($2\Theta=62.75°$), and the Φ scan is of the (101) rutile plane ($2\Theta=36.08°$, $\chi=32.79°$). For the (100) orientation, the Θ scan is of the (200) rutile peak ($2\Theta=39.19°$), and the Φ scan is of the (110) rutile plane ($2\Theta=27.45°$, $\chi=45.00°$).

Using the Θ and Φ FWHM's as an indication of crystallographic microstructural quality, the table shows a number of doping effects. First, nominal 0.5 at% Ga-doping of the (001) rutile causes an initial 44% increase in the Θ scan FWHM value, then remains constant to 4.5 at% Ga. Under the same conditions, the Φ scan FWHM value continually increases to 75% greater at 4.5 at% Ga. Second, 0.5 at% Ga-doping of the (100) films causes a 13% decrease in Θ. The FWHM values for the Φ peaks are not reported because they are very broad due to the three possible in-plane orientations of the tetragonal rutile lattice on the three-fold symmetry of the (0001) sapphire surface. Finally, Nb-doping of the (001) rutile causes continual increases in Θ and Φ FWHM, with values to 41% and 42% greater at 4.5 at% Nb compared to the respective undoped values.

Table I. Doped Heteroepitaxial Thin Films Microstructure Summary

Film	Substrate	Dopant (at %)	Θ FWHM (degrees)	Φ FWHM (degrees)	RMS Roughness
(001) Rutile	($10\bar{1}0$) Al_2O_3	None	0.34	1.06	3.4 Å
[100] // [$11\bar{2}0$] (11.5%)*		0.5 Ga	0.49	1.46	--
[010] // [0001] (6.0%)*		1.0 Ga	0.50	1.60	17.6 Å
		2.5 Ga	0.47	1.67	--
		4.5 Ga	0.46	1.85	7.5 Å
		1.0 Nb	0.38	1.24	5.3 Å
		4.5 Nb	0.48	1.50	6.6 Å
(100) Rutile	(0001) Al_2O_3	None	0.47	--	7.0 Å
[010] // [$\bar{2}110$] (-3.5%)*		0.5 Ga	0.41	--	--
[001] // [$01\bar{1}0$] (-3.5%)*					

Growth Conditions: Background gas partial pressure (Torr): 0.8 He, 0.8 O_2; Growth Temperature: 650°C; Growth Rate: 40 Å/min; Thickness: 3500Å   * Cationic mismatch for undoped films

The (001) X-ray data trends can be explained by analysis of the dopant incorporation into the lattice. The lattice mismatch data mentioned above are given as positive values because this orientation of rutile is in compression in each surface direction on sapphire. When Ga ions are substituted into the lattice, they are ionically compensated by Ti interstitials [9-10] which further expand the lattice already under compression. This increased lattice expansion deteriorates the epitaxial quality of the film. Nb is electronically compensated but is a larger ion than Ti, which also expands the lattice and decreases the epitaxial quality as shown by the increasing Θ and Φ FWHM values. The (100) lattice mismatch is reported as negative because this rutile orientation is under tension. The lattice expansion caused by ionic compensation of Ga improves the epitaxy as shown by the decrease in the Θ value.

The final column lists the RMS roughness as determined by AFM. Doping of the (001) oriented films with both Ga and Nb increases the surface roughness. The same surface morphology in terms of feature size and shape is, however, generally maintained. The (100) oriented undoped film is slightly rougher than the (001) film.

Figure 1 shows SIMS depth profiles of annealed and as-grown (001) and (100) oriented 1.0 at% Ga-doped films. The data reveal that segregation occurs nominally within the first 100 Å for both orientations in as-grown or annealed samples. The segregation profile can be modified by annealing, however, in this study we have examined only as-grown films. Dopant segregation will create space charge regions which are important to the electrical characterization described below.

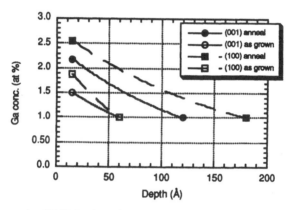

Fig. 1 Ga concentration SIMS depth profiles for (001) and (100) oriented 1% Ga-doped films.

Figure 2 is a plot of the logarithm of the inverse resistance (R) versus inverse absolute temperature of the Nb-doped, undoped and Ga-doped (001) oriented films. These data were obtained in a dry N_2 atmosphere. At constant temperature, the conductance (1/R) of the 2 at% Nb-doped film is greater than that of the undoped film, and both films have conductances which are greater than the Ga-doped films. Above 1 at% the conductance of the Ga-doped films is independent of doping level. The activation energy for conduction of the Nb-doped film is similar to that of the undoped film (~0.1 eV). The activation energy for conduction of the Ga-doped films is much greater (~0.8 eV) which indicates that these films have a different conduction mechanism.

TiO_2 is an n-type semiconductor [11,12]. Upon doping with Nb, an electronically compensated donor, more electrons are introduced into the lattice, which causes the observed increase in conductance over the undoped film. The similar activation energies indicate a similar conduction mechanism. Ga-doping is ionically compensated by Ti interstitials which introduces defects into the lattice. These defects may act as traps which decrease the conductance and produce the change in conduction mechanism.

227

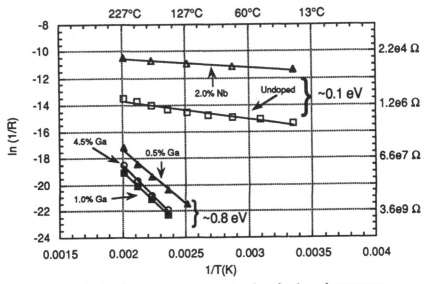

Fig. 2 Arrhenius plot of conductance as a function of reciprocal temperature.

Figures 3 and 4 are plots of the normalized resistance of several films exposed to dry and humid N_2 and air atmospheres, respectively. Each major tick of the y-axis represents 20% of the total resistance value measured under dry conditions and shown in ohms at the top of each bar. All measurements were taken at 225°C in the [100] surface direction. Similar humidity effects are also observed in the [010] direction. In N_2 (Fig. 3), humidity has the largest effect on the resistance of the Ga-doped films. The decrease is greater than a factor of four for 0.5% Ga and increases to an order-of-magnitude for 1.0% Ga. Humidity causes a small decrease in the resistance of the Nb-doped and undoped films. Figure 4 shows normalized resistance of the same films shown in Fig. 3, but now in an air atmosphere. As expected for an n-type semiconductor, all the resistance values increase with an increase in oxygen partial pressure. The humidity effects, however, are similar to those observed in N_2. Little or no effect is observed on the resistance of the Nb- and undoped films with the addition of humidity, but the resistance of both 0.5 and 1.0 at% Ga-doped films decreases by a factor of two with added humidity.

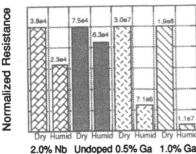

Fig. 3 Humidity effects on resistance (Ω) for (001) TiO_2 in N_2.

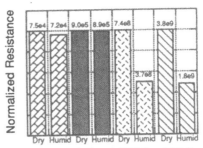

Fig. 4 Humidity effects on resistance (Ω) for (001) TiO_2 in air.

The differences in the effect of humidity on the resistance appears to correlate with the differences in conduction mechanism. The effectiveness of Ga-doping in increasing the sensitivity of rutile to humidity may be due to the presence of excess Ti atoms or O vacancies at the surface [10,13]. These surface sites have been suggested by a number of researchers as the preferred site for water adsorption on the TiO_2 surface [14-17].

To determine the depth into the film that is affected by humidity, undoped and 1.0 at% Ga-doped films of 150 Å thickness were grown. Figure 5 shows a comparison of the humidity effect on the resistance of these thin films in a N_2 atmosphere compared to 3500 Å films of the same composition. The humidity effect for both Ga-doped films was similar, indicating that the conduction and the changes with humidity in Ga-doped films must occur within the first 150 Å below the surface. The resistance of the thin undoped film, however, exhibits a similar decrease in the presence of humidity to the Ga-doped films. Therefore, when the conduction path in the undoped films is confined to the near surface region, it can be similarly affected by humidity.

To investigate the role that proton migration into the films plays in the effect of humidity on resistance, a 0.5 at% Ga-doped (100) oriented film was grown and tested. The activation energy for proton migration was reported to be a factor of two greater perpendicular to the c-axis as opposed to parallel [18]. Figure 6 shows that the effect of humidity on the resistance of these films is similar. Therefore, proton migration does not appear to significantly participate in the observed humidity effect.

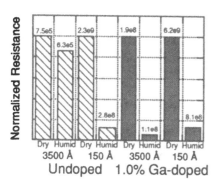

Fig. 5   Humidity effects on thick and thin Ga- and undoped films

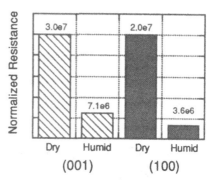

Fig. 6   Humidity effects parallel and perpendicular to the c-axis for 0.5 at% Ga-doped films

SUMMARY

Our investigation revealed a number of trends in the microstructural and electrical properties of heteroepitaxial rutile thin films and their relation to the effects of humidity on these films. The epitaxial quality of the films was observed to decrease with increasing Ga and Nb doping levels, as evidenced by increasing values of the FWHM's of the dominant diffraction peaks in Θ and Φ scans. In dry atmospheres at constant temperature, the conductance of Nb-doped films is greater than that of undoped films which is greater than that of Ga-doped films. Also in dry atmospheres, the activation energy for conduction in Nb-doped films is less than or equal to the activation energy of undoped films (~0.1 eV). These were found to be much less than the activation energies of Ga-doped films (~0.8 eV). Humidity effects are enhanced in a N_2 versus air atmosphere and are found to be greatest for Ga-doped and thin (~150 Å) undoped films. Proton migration does not appear to play a significant role in the effects of humidity on the electrical conduction in rutile films.

REFERENCES

1. J.G. Fagan and V.R.W. Amarakoon, Am. Cer. Soc. Bul. **72**, p. 119 (1993).

2. B.M. Kulwicki, J. Am. Cer. Soc. **74**, p. 697 (1991).

3. S.P. Lee, J.Y. Rim and Y.K. Yoon, Sens. and Mat. **7**, p. 23 (1995).

4. J. Pennewiss and B. Hoffman, Mat. Let. **5**, p. 121 (1987).

5. M. Radecka and M. Rekas, J. Phys. Chem. Sol. **56**, p. 1031 (1995).

6. K. Katayama, K. Hasegawa, Y. Takahashi, and T. Akiba, Sens. and Act. A **24**, p. 55 (1990).

7. J.-H. Park and S.J. Park, J. Mat. Sci.: Mat. in Elec. **5**, p. 300 (1994).

8. D.R. Burgess, P.A. Morris Hotsenpiller, T.J. Anderson and J.L. Hohman, J. Crystal Growth **166**, p. 763 (1996).

9. J. Yahia, Phys. Rev. **130**, p. 1711 (1963).

10. J.S. Ikeda and Y.-M. Chiang, J. Amer. Cer. Soc. **76**, p. 2437 (1993).

11. A. Takami, Am. Cer. Soc. Bul. **67**, p. 1956 (1988).

12. E.C. Subbarao, Ferroelectrics **102**, p. 267 (1990).

13. J.S. Ikeda and Y.-M. Chiang, J. Amer. Cer. Soc. **76**, p. 2447 (1993).

14. W. Göpel, U. Kirner and H.D. Wiemhöfer, Sol. St. Ionics **28-30**, p. 1423 (1988).

15. K. Sakamaki, K. Itoh, A. Fujishima and Y. Gohshi, J. Vac. Sci. Tech. **A8**, p. 614 (1989).

16. H.O. Finklea, Semiconductor Electrodes, Elsevier, Amsterdam, 1988, p. 519.

17. W.J. Lo, Y.W. Chung and G.A. Somorjai, Surf. Sci. **71**, p. 199 (1978).

18. O.W. Johnson, S.-A. Paek and J.W. DeFord, J. Appl. Phys. **46**, p. 1026 (1975).

Molecular Modeling of Selective Adsorption from Mixtures

T. J. BANDOSZ[‡], F. J. BLAS[¶], K. E. GUBBINS[§], C. L. McCALLUM[§*],
S. C. McGROTHER[§], S. L. SOWERS[§†] and L. F. VEGA[¶]
‡ Department of Chemistry, The City College of New York, New York, NY 10031
§School of Chemical Engineering, Cornell University Ithaca, New York, 14853
¶ Escola Tècnica Superior d'Enginyeria Quìmica, Carretera de Salou, s/n 43006 Tarragona, Spain

ABSTRACT

Molecular simulation methods provide a means for carrying out systematic studies of the factors affecting adsorption phenomena. For selective adsorption, the selectivity is strongly affected by the interaction energy with the pore walls, molecular size and shape, site specific interactions, entropic effects, differences in diffusion rates, and networking effects. Two recent studies of site specific selectivity will be described. The first is an investigation of the effect of oxygenated surface sites on the adsorption of water vapor on activated carbons. Hydrogen-bonding sites are modeled using off-center square well interactions for both water and wall sites; wall sites are placed at the edges of the graphite micro-crystals. New experimental results for water adsorption at low pressures on carefully characterized activated carbons are reported, and are found to be in good agreement with the simulations. In the second application, we consider the separation of alkene/alkane mixtures using aluminas whose surfaces have been doped with metal ions. π-complexation between these metal ions and the alkenes can produce a highly selective separation. The simulations are found to be in good agreement with the available experimental data, and have been used to predict separations for other conditions not yet studied in the laboratory.

INTRODUCTION

The adsorption of water on activated carbons is qualitatively different from that of simple fluids. There are two sources of this dissimilarity: (a) the water-water interaction is very strongly attractive, compared to simple fluids, and (b) the adsorption of water is largely controlled by the formation of H-bonds with oxygenated groups on the surface. For simple fluids, pore filling occurs via the formation of a fluid monolayer on each surface, often followed by a second and further layers, prior to capillary condensation and pore filling. By contrast, water molecules adsorb onto oxygenated surface sites and these adsorbed molecules provide nucleation sites for the formation of larger water clusters; eventually these clusters connect, and pore filling occurs. When the density of oxygenated sites on the surface is appreciable, the pore filling seems to occur by a continuous filling process without capillary condensation.

The modeling of water on graphitic carbons is straightforward. The absence of oxygenated sites on the surfaces permits even simple representations of the system to mimic the observed type V isotherms [1]. In such isotherms, there is very little adsorption at moderate pressures, and pore filling occurs rapidly at high relative pressure. Both point charge and square well models of water reproduce the phase behavior well [2]-[6]; moreover even simple Lennard-Jones models capture the essence of the isotherm, if the fluid-wall interaction is sufficiently weak compared to the fluid-fluid potential [7].

When the surface activation is significant, the adsorption isotherm is more complex and the models needed to reproduce the observed behavior are consequently more involved. Segarra and Glandt [8] examined the simple point charge (SPC) model of water and a pore structure comprising a series of randomly oriented platelets, around the edges of which dipolar forces act. This model produced isotherms qualitatively in accord with experimental results. However the lack of distinct surface groups is thought to be a major short-coming of this model. Maddox, Ulberg and Gubbins [5] explicitly accounted for discrete surface sites, using the OPLS model to represent carbonyl groups on the carbon walls (such a model

---

*current address: Intel Corporation, 5000 W. Chandler Blvd, Chandler, AZ 85226.
†current address: Westvaco, Laurel Technical Center, 1101 Johns Hopkins Rd, Laurel, MD 20723.

contains both Lennard-Jones and Coulombic terms). The water model used by these authors also contained point charges (TIP4P). This study also yielded adsorption isotherms in qualitative agreement with experimental observation. The long-range nature of the Coulombic potentials makes such simulations expensive, and the usual methods used to account for the long tail in the potential are inappropriate for confined geometries. Hence, these authors used simple minimum image convention (MIC) and assumed that their system size was sufficiently large to prevent systematic errors. Similar results were obtained by Müller et al. [6] using a simpler model. In this study, the water comprised a spherical Lennard-Jones core with 4 embedded square well sites tetrahedrally arranged (2 for the hydrogens and 2 for the lone-pair electrons), to mimic the hydrogen bonding of water. Since all the elements of this model act over only short ranges, the truncation problems of other models are avoided. The activated carbon walls consist of square-well sites rigidly attached to a structureless 10–4–3 wall [9]. The model of Müller et al. not only predicts the adsorption isotherm as accurately as the model of Maddox et al., but (with careful fitting of parameters) it also reproduces the bulk structure and phase behavior of fluid water with surprising accuracy. The model's simplicity allows a greater amount of the available parameter space to be studied, and is therefore the model studied herein.

Olefin/paraffin separations play a central role in the chemical and petrochemical industries. Such mixtures usually result from the thermal or catalytic cracking of hydrocarbons, and the majority of them contain ethylene as a coproduct. Due to the small difference in volatility of these compounds, traditional separation processes, such as fractional distillation, are very expensive, especially at the end of the process or at very low concentrations of one of the components. For these reasons, during recent years, many groups have been looking for new processes or designing new materials to separate this kind of mixture. One of the most promising processes is adsorption via π-complexation.

The method is based on new adsorbents synthesized by spontaneous dispersion of a monolayer of CuCl on $\gamma-Al_2O_3$ [10, 11]. The Cu(I) ions are able to form complexes with the double bonds of the olefins. The main advantage is that the adsorption can be reversed by changing the pressure and/or the temperature, making it easy to recover the material from the porous adsorbent.

In this study we use a recently proposed model [12], based on simplified intermolecular potentials, to quantitatively predict the adsorption behavior of these fluids and their mixtures. We obtain the single component adsorption isotherms of ethane and ethylene on bare $\gamma-Al_2O_3$ and on $CuCl/\gamma-Al_2O_3$, and compare these predictions with the experimental results. The agreement between experimental data and simulation results is very good, and we are able to predict the adsorption behavior of this mixture at different thermodynamic conditions.

EXPERIMENTAL: Water on activated carbon

The adsorbent was Norit activated carbon N2 (Sorbonorit 2–A5998). It was oxidized with 30% hydrogen peroxide at 323K for 2 hours. The details of this procedure are described in references [13, 14]. The nitrogen adsorption isotherm was measured by an ASAP 2010 (Micrometrics) sorptometer at 77K. Before the experiment the sample was heated for 10 hours at 393K and then out-gassed under a vacuum of 10^{-6} Torr. This isotherm was used to calculate the specific surface area, micropore volume, and pore size distribution (PSD) using density functional theory (DFT). The density of oxygen-containing surface sites is determined by Boehm and potentiometric titration; a value of 0.675 sites/nm^2 is obtained. The sorption isotherm of water was obtained at 298K using the same ASAP 2010 (Micrometrics) instrument equipped with a vapor sorption kit. Prior to the experiment the sample was out-gassed at 393K and 10^{-6} Torr. Dissolved gases were removed from the water by a series of melting/freezing cycles with out-gassing of the evolved contaminants. Care was taken to perform measurements to very low relative pressures.

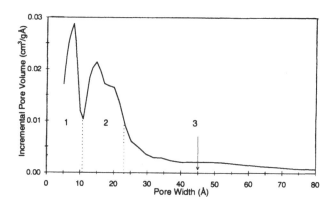

Figure 1: Experimental pore size distribution as determined by DFT for N2 activated carbon.

MODELS: Water on activated carbon
<u>Pore Size Distribution of activated carbon</u>

The active carbon used in our experimental work has a distribution of pore sizes and shapes; the pore-size distribution, obtained by density functional theory (DFT), is displayed in figure 1. Clearly there is a bimodal distribution of pore-widths, and a long tail in the meso-porous range. Note that the PSD from experiment gives information on *effective* pore widths. This quantity can be defined as:

$$w = H - \sigma_c \qquad (1)$$

where H is the actual pore-width, and σ_c can be thought of as an effective carbon diameter. The amount of adsorbate adsorbed at a given pressure can be approximated as:

$$N = \int_0^\infty f(w) N(w) \, dw, \qquad (2)$$

where w is the effective pore width defined above, $f(w)$ is the PSD and $N(w)$ is the amount adsorbed in pores of this width, at this pressure. Obviously simulating a continuous distribution of pore sizes, as represented by this equation, is impossible. However the integral can be approximated as a finite sum over discrete pore-widths:

$$N = c_1 N(w_1) + c_2 N(w_2) + \ldots + c_n N(w_n) \qquad (3)$$

where w_i are the selected pore-widths, and c_i are weighting factors indicative of the total pore volume represented by the individual w_i's. For simplicity we assume that all pores are slit-shaped, and that networking effects are negligible. Since the simulations are very time-consuming we choose 3 pore-widths, w_i: 0.99, 1.69 and 4.5nm. The weighting factors c_i for these three pore sizes are: 0.518, 0.371 and 0.111, respectively. These factors are determined by the area under the three portions of the curve.

Equation (3) is employed only for our low pressure analysis ($P/P_0 \leq 0.02$). The method breaks down in the pore-filling region, since pores of different width fill at different pressures. Weighted averaging necessarily leads to a stepped isotherm, unless a large number of pore widths are simulated. Thus a more general procedure was sought to recreate the full adsorption isotherm.

For larger pressures we create a single smooth adsorption isotherm from our individual pore simulations by fitting the numerical data to an empirical equation for a water/active carbon system. The equation of Talu and Meunier [15] was selected, since it was developed specifically for associating systems. Their equation can be written:

$$P = \frac{H\Psi}{1 + K\Psi} \exp\left(\Psi/N_m\right) \tag{4}$$

where K is a parameter related to the equilibrium constant for dimer formation, N_m is the saturation capacity of the pore (and thus varies with pore-width), $H = \lim_{P \to 0} dP/dN$ is Henry's constant, and

$$\Psi = \frac{-1 + (1 + 4K\xi)^{1/2}}{2K} \tag{5}$$

and

$$\xi = \frac{N_m N}{N_m - N} \tag{6}$$

with N being the amount adsorbed. Thus our fitting procedure yields values for H, K, N_m, and a continuous isotherm for each of the pore-widths studied. We interpolate smoothly between these fitting parameters to estimate their values for pore-widths other than those simulated. Using these interpolated fitting parameters, we can construct an adsorption isotherm for any pore-width. Finally we generate a single adsorption isotherm by averaging the adsorption isotherms for many pore widths using equation (2).

Surface Groups

The adsorption of water depends strongly upon the surface chemistry of the active carbon. Not only the density and arrangement of sites, but also the nature of the surface groups will lead to differences in the amount adsorbed at a given pressure. X-ray diffraction studies have indicated that oxygen-containing surface groups, are located around the edges of carbon micro-crystallites [16]. Groups such as carboxylic, phenolic, carbonyl, lactonic, chromene and pyrone have been identified on active carbon surfaces. The relative and absolute amounts of these groups depend upon the activation conditions employed in the creation of the porous sample. The chemically distinct surface groups will have somewhat different interaction energies with the adsorbate, but for computational simplicity, we assume all sites are equivalent. Thus all sites have the same water/site binding energy in our model and each site can bond to only one water molecule.

Surface groups are known to be attached at the edges of the carbon micro-crystals which make up the carbon walls. In our simulations the surface sites are placed on a square grid. There is no clear consensus as to what typical microcrystal sizes should be; values between 1 and 10nm are reported for the length of a crystal edge. Arbitrarily, we select $L = 2$nm as our crystallite length for all the adsorption simulations reported here. The same arrangement of sites is maintained for the simulations at different pore-widths.

Intermolecular Potentials

Water is modeled as a Lennard-Jones (LJ) sphere, with four tetrahedrally arranged off-center square-well sites to mimic H-bonding sites (see figure 2). Two sites represent hydrogen atoms, and two mimic the lone-pairs of electrons. Bonding only occurs between unlike sites. The position and range of the sites are fixed ($0.42\sigma_{LJ}$ from the center of the sphere, site diameter $\sigma_{hb} = 0.0612$nm), to reproduce typical O–O separations for liquid water [17]. Isolated water molecules are known to deviate from perfect tetrahedral geometry (the H–O–H bond angle is considerably less than 109.5°), however, in the H-bonded liquid structure, this effect is known to diminish, and the water molecules adopt very nearly perfect tetrahedral bond angles. Thus fixing the site positions should be a fair assumption. The LJ parameters are taken as in previous studies: $\sigma_{ff} = 0.306$nm, $\epsilon_{ff} = 90kT$. This only leaves the strength of the water-water H-bond. We carefully fit this parameter to best reproduce bulk behavior. We find a value of $\epsilon_{hb} = 3800kT$ gives the best fit to bulk vapor density, vapor pressure and coexisting liquid density at 298K: all three of these properties are obtained to within 2%.

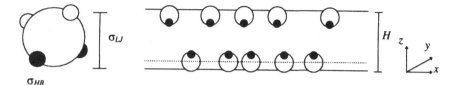

Figure 2: Schematic of model water molecule used in simulation.

Figure 3: Active carbon slit pore model. Lines represent the pore walls, separated by a distance H. Large circles are LJ oxygen, small circles are SW hydrogen atoms.

For the interaction of the fluid with the wall, the carbon is modeled by the structureless 10–4–3 potential [9],

$$\phi(z) = 2\pi \rho_{ss}\epsilon_{sf}\sigma_{sf}^2 \Delta \left[\frac{2}{5} \left(\frac{\sigma_{sf}}{z} \right)^{10} - \left(\frac{\sigma_{sf}}{z} \right)^4 - \left(\frac{\sigma_{sf}^4}{3\Delta \left(z + 0.61\Delta \right)^3} \right) \right] \tag{7}$$

with the density of the carbon $\rho_{ss} = 114\text{nm}^{-3}$, the spacing of graphite planes $\Delta = 0.335\text{nm}$, the solid-solid potential well-depth $\epsilon_{ss}/kT = 28$ and the carbon diameter $\sigma_{ss} = 0.340\text{nm}$. Fluid-solid cross parameters are calculated using the Lorentz-Berthelot rules: the diameter is the arithmetic, and the well depth the geometric mean of the pure component values,

$$\sigma_{sf} = (\sigma_{ss} + \sigma_{ff})/2 \tag{8}$$

$$\epsilon_{sf} = (\epsilon_{ss}\epsilon_{ff})^{1/2}. \tag{9}$$

The active surface sites consist of LJ spheres with a single bonding site; these represent hydroxyl groups. The LJ sphere is the same size as that of the water molecule, and is centered 0.1364nm above the wall (see figure 3). The site is embedded $0.42\sigma_{LJ}$ from the center of the LJ sphere, and is placed as far from the wall as possible, i.e., on the normal to the wall which passes through the center of the LJ sphere. The wall-site square well has the same range as those on the water molecules, and is only able to form bonds with the lone-pairs on the fluid molecules. The effect of site density, microcrystal size, and surface site arrangement will be the subject of a future study [18].

MODELS: Ethylene/ethane mixtures on Alumina

Pore Model

We have modeled alumina as a mono-disperse porous solid with pores of cylindrical shape (see figure 4). The pore diameter $D = 1.10\text{nm}$ is chosen to fit the shape of the experimental adsorption curve at low coverages. The oxygen ions are modeled as Lennard-Jones (LJ) sites, with potential parameter values taken from the literature [20].

The presence of the CuCl molecules on the activated surface has two major effects: a decrease of the available space for the adsorbate molecules and a change of the effective dispersive interaction between the adsorbate molecules and the wall. Since the alumina surface is covered by a full monolayer of CuCl, we can expect the reduction in the mean pore diameter to be of the order $\sigma_{Cu} + \sigma_{Cl}$, where these are the diameters of the Cu and Cl ions. This suggests an effective mean pore diameter for the CuCl/γ−Al$_2$O$_3$ of about 0.75–0.85nm. Comparison of simulations of this model with experimental data [10] for the adsorption of pure ethane and ethylene on this material show that $D = 0.80\text{nm}$ is a good choice. This is consistent with our expectations, based on the mean diameters of Cu$^+$ and Cl$^-$ ions, and we have adopted this pore diameter for CuCl/γ−Al$_2$O$_3$ in all of calculations involving the activated surface.

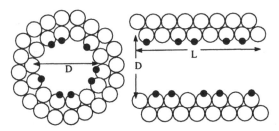

Figure 4: Schematic of the cylindrical pore used in this work to model CuCl/$\gamma-$Al$_2$O$_3$. Small circles represent Cu ions placed at random locations on the surface (with the constraint that no oxygen can have more than one SW site). The bare $\gamma-$Al$_2$O$_3$ surface is identical except for the removal of surface sites.

Fluid-fluid potentials

We have described the molecules as made up of united atoms, i.e., CH$_3$ and CH$_2$ are considered as single interaction units. The interaction between two units i and j on different molecules is described by the Lennard-Jones potential with the molecular parameters given by Jorgensen et al. [19]. These molecular parameters, ϵ_{ij} and σ_{ij}, represent the dispersive energy and the diameter of the sites. An additional parameter, the bond length, is used to account for the geometry of the molecules, L_b, the center-to-center distance between the two sites of every molecule. For ethane these values are $\sigma = 0.3775$nm, $\epsilon/k_B = 104.15$K, $L_b = 0.153$nm; and for ethylene: $\sigma = 0.385$nm, $\epsilon/k_B = 70.44$K and $L_b = 0.134$nm.

The model proposed [12] mimics the π-complexation interaction between ethylene molecules and Cu ions on the surfaces through two square well (SW) associating sites placed in the line perpendicular to the symmetry axis of the molecule, and at distance of $0.5\sigma_{\text{ff}}$ from the center of mass, where ff refers to the fluid-fluid parameters of ethylene molecules. These sites are characterized by σ_π and ϵ_π, the size and energy parameters of the SW sites and r_{AB}, the site-site distance. The site diameter is chosen equal to $0.2\ \sigma_{\text{sf}}$. The well depth ϵ_π is calculated from the experimental heat of adsorption at low coverages from the experiments [10]. The value found is $\epsilon_\pi = 1400$K. Note that the SW sites have no effect on the fluid-fluid interactions. The Lorentz-Berthelot combining rules are again used to described the intermolecular interactions between unlike molecules.

Fluid-solid potentials

The interaction between the molecules and $\gamma-$Al$_2$O$_3$ surface is a sum over all the LJ terms, characterized by σ_{sf} and ϵ_{sf}, where $\sigma_{\text{sf}} = (\sigma_{\text{ss}} + \sigma_{\text{ff}})/2 = 0.303$nm. The dispersive energy parameters between fluid molecules and oxygen ions (ϵ_{sf}) are fitted to the low pressure adsorption data and, for the $\gamma-$Al$_2$O$_3$ surface, are found to be 95 and 98K for ethane and ethylene, respectively [12]. The interactions between the adsorbate molecules and the Al^{3+} ions are not accounted for explicitly. However, their presence, is effectively accounted for in the LJ parameter values.

In the case of CuCl/$\gamma-$Al$_2$O$_3$ interaction between the adsorbate molecules and the Cl$^-$ ions is taken into account by reducing the dispersive energy values of ethane-oxygen and ethylene-oxygen to 48.5 and 87K, respectively. These values are found by fitting to low pressure adsorption data for ethane and ethylene on CuCl/$\gamma-$Al$_2$O$_3$ [10]. The structured pore surface of CuCl/$\gamma-$Al$_2$O$_3$, with a nearly monolayer of Cu$^+$ ions over the surface, is modeled by placing SW associating sites on the oxygen ions in the surface layer of the cylinder. The sites are randomly placed on the surface, at a distance $0.5\sigma_{\text{ss}} = 0.1515$nm from the wall,

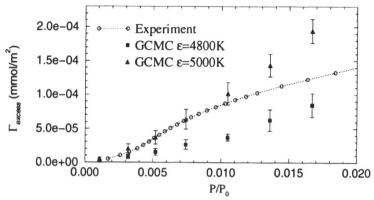

Figure 5: Low pressure adsorption isotherm of water on activated carbon at 298K.

and have identical size and energy parameters as those used for the ethylene molecule. The site density is found from the experimental spreading of CuCl over $\gamma-Al_2O_3$ [21].

MOLECULAR SIMULATION METHOD

We have used Monte Carlo simulations in the grand-canonical (μVT) ensemble, with random creation and destruction of particles and standard Metropolis Monte Carlo translation/rotation moves [22]. Each of these three moves is attempted equally often: the type of move is chosen randomly. The maximum rotation and displacement are altered so that the combined move has a $\approx 40\%$ probability of acceptance. It is noteworthy that the maximum displacement/rotation parameters are very small compared to values encountered for non-associating systems.

The simulation cell is 4nm (10nm in the low pressure studies) in the directions parallel to the walls and periodic boundary conditions and minimum image convention are applied in these directions. Such conditions should ensure that finite size effects can be ignored.

The adsorption isotherm commences with the cell empty, and a value of the fugacity corresponding to a low pressure is input. The final configuration generated at each stage is used as the starting point for simulations at higher fugacities. The desorption isotherm begins with a filled pore and the fugacity is reduced in a step-wise fashion, with final configurations once again used as starting points for subsequent simulations. For the case of mixtures, the type of molecule to be moved is randomly chosen with a probability equal to 0.5. Also, the probability of creation/deletion of a type of molecule is 0.5.

The calculated second virial coefficients of ethane and ethylene show that non-ideal gas contributions to the pressure are, in all cases, less than 1%. Thus the ideal gas relationship between the absolute activity of species i and the total bulk pressure of the fluid could be used to calculate the bulk gas pressure.

RESULTS: Water on Activated Carbon

Low pressure region

In figure 5 we show the low pressure adsorption isotherm of water into the activated carbon at 298K. The simulation data is for two wall-fluid H-bonding strengths $\epsilon = 4800$ and $\epsilon = 5000$K. The points are the weighted average of the three pore sizes outlined above. The simulation results for $\epsilon = 5000$K show excellent agreement with experiment at the lower pressures, up to $P/P_0 \approx 0.001$, but are too high for the higher pressures. The results for $\epsilon = 4800$K, while too low for the low pressures shown in figure 5, give the best agreement over the entire pressure range (see figure 6). There is a distinct 'S'-shape to the experimental data which leads us to conclude that the mechanism of water adsorption at

237

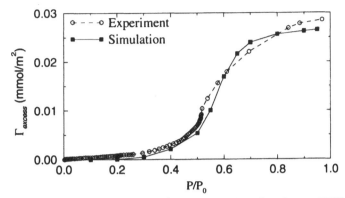

Figure 6: Adsorption isotherm of water on activated carbon at 298K.

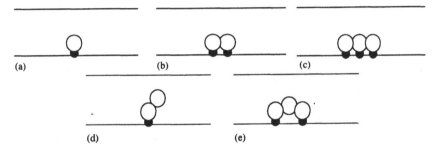

Figure 7: Arrangements of water molecules (open circles) at low pressures: (a) a single site-adsorbed water molecule, (b) two wall site-adsorbed molecules bonded to each other, (c) three wall site-adsorbed molecules bonded to each other, (d) a molecule bonded to a site-adsorbed water molecule, (e) a molecule bonded to two site-adsorbed water molecules.

these low pressures is more complex than previously envisaged: at the lowest pressures, only direct bonding of water with surface sites occurs (fig 7A). As soon as some water is adsorbed, certain sites become more attractive as water can adsorb onto these sites and simultaneously bind to previously adsorbed water, i.e., cooperative bonding (fig 7B, 7C); in this region the slope of the isotherm increases. Such sites are quickly exhausted and only water-water bonds may now form (fig 7D, 7E), and the slope decreases. This mechanism is borne out by simulation snapshots, which show exactly this pattern of adsorption with increasing pressure. The simulation data captures this 'S' shape, particularly for $\epsilon = 5000K$ but the later decrease in slope occurs at too high a pressure. We attribute this failing to our assumption that all sites have the same strength and form only one bond. In real surface groups, the cooperative bonding will be most prevalent between H-bonding sites within the same chemical group. The sites in our model are all identical, and the cooperative bonding occurs only when two such sites are, by chance, in close proximity. Allowing surface groups to bond to more than one water, or incorporating energetic inhomogeneity, should make the model more realistic.

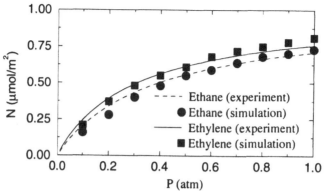

Figure 8: Single component adsorption isotherms for ethane and ethylene on $\gamma-Al_2O_3$ at T=333K. Symbols represent simulation results for ethane (circles) and ethylene (squares). Lines are correlations from experimental [11] data for ethane and ethylene.

Entire pressure range

We have performed MC simulations for four pore-widths (the three used at low pressures and a very narrow pore of 0.79nm). Isotherms for other pore-widths are obtained by interpolation (using the Talu-Meunier equation as described in the model section). The resultant isotherm (figure 6) is in reasonable quantitative accord with the experimental results. Both curves have the expected type V shape: very little adsorption occurs up to some pressure where there is rapid uptake, followed by a flat region corresponding to filled pores. There are however several discrepancies: at low pressures the simulated data lie below the experimental curve; also while the overall filling pressure is in good agreement, the simulated curve is steeper in this region, and finally the filled pore capacity is underestimated by the numerical data. We address each of these points here. The under-estimate of the adsorption at low pressure is due to the fitting of the simulated data to the Talu and Meunier [15] equation (equation 4). For the narrowest pore studied, there is considerable adsorption prior to pore filling, but the best fit to the equation of state fails to reproduce this. Thus an improved equation should better reproduce the low pressure data. There are several potential causes of the difference in slope at the filling pressure. The fits to the equation generally appear to be steeper in the filling region than the simulation points to which they are applied; also there is some uncertainty in the PSD, which would show up most clearly in this region of the isotherm. Again an improved equation should ameliorate this region of the curve. Uncertainty in the PSD causes the erroneous pore capacity. None of the disparities between the experimental and numerical curves are directly attributable to failings in the model, although these of course may exacerbate the inconsistencies, particularly at low pressures.

RESULTS: Ethylene/ethane mixtures on Alumina

Single component adsorption isotherms

In figure 8, we show the adsorption isotherms of ethane and ethylene on bare $\gamma-Al_2O_3$; it can be seen that bare alumina is not able to separate these components. The adsorption of both is quite similar. The agreement between experimental data [11] and molecular simulation results is excellent in all cases.

In figure 9 is shown the adsorption isotherms on $CuCl/\gamma-Al_2O_3$. Under the same thermodynamic conditions as for $\gamma-Al_2O_3$, the adsorption of ethylene is greatly increased, while the adsorption of ethane slightly decreases with respect to the bare alumina. Again,

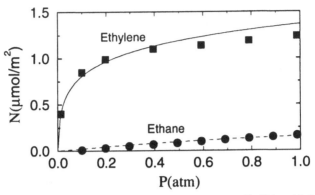

Figure 9: Adsorption isotherms for ethane and ethylene on CuCl/$\gamma-$Al$_2$O$_3$ at T=333K. Symbols represent simulation results for ethane (circles) and ethylene (squares). Lines are correlations from experimental data [11] for ethane and ethylene.

predictions from the model are in excellent agreement with experiment. Small deviations between the predicted adsorption of ethylene at higher pressures and that obtained experimentally are observed; pore filling occurs earlier for the model than is observed experimentally. This is probably due to the fact that we are modeling a single pore size, while the actual material has a pore size distribution. Although our model pore represents the average pore size of the alumina, in the experiments there are some pores with a greater diameter, allowing more ethylene molecules to fit inside.

Selectivity of ethylene

Since the molecular model describes pure component adsorption well, it can be used to predict the adsorption behavior of the mixture. No experimental data exists for the mixture. However, agreement between experiments and simulations is expected, since the molecular parameters do not depend on the thermodynamic conditions of the system.

The selectivity of ethylene over ethane as a function of pressure is shown in figure 10 at two different temperatures, 300 and 333K, for a bulk composition of 0.9 for the ethane mole fraction.

We observe that, even at the higher temperature studied, the activated alumina is strongly selective for ethylene. The selectivity is multiplied by a factor of two when the temperature is decreased. As expected, selectivity decreases when the pressure is increased, due to partial pore filling.

CONCLUSIONS

A detailed simulation study has been undertaken in order to determine the underlying mechanism of water adsorption into active carbon pores, and to reproduce the adsorption isotherm of water into a well-characterized activated carbon material. The low pressure adsorption data provides information concerning the mechanism of water adsorption. The 'S'-shaped isotherm suggests that cooperative bonding occurs after a very narrow initial stage where only direct fluid-wall binding is possible. A second inflexion indicates that there is a limited number of surface positions where such cooperative bonding may occur, and once these positions are exhausted the steep rise in adsorption ceases.

We conclude that the molecular model used must satisfy two criteria in order to reproduce correct, overall behavior: the fluid potential must accurately predict the liquid structure and density at coexistence; and the pore-fluid interaction must reproduce the very low pressure adsorption isotherm data. The model used herein certainly reproduces the liquid phase with sufficient accuracy. However, the fluid-wall potential is seen to lack some important features

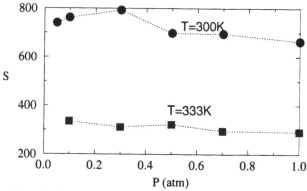

Figure 10: Selectivity of ethylene over ethane on CuCl/$\gamma-Al_2O_3$ at 300K (circles) and 333K (squares). The mole fraction in the bulk phase is $y_{ethane}=0.90$.

and this leads to qualitative discrepancies in the very low pressure region. These failings we attribute to the energetic and steric homogeneity we impose upon our wall-fluid bonding. In reality the binding sites will be individual components of complex surface groups, the precise chemical and physical nature of which may lead to a large inhomogeneity in the interaction energy of water with the wall sites.

There is a further requirement of the numerical study, which is that the simulation results for various pore sizes, must be somehow 'patched' together, using the experimentally obtained PSD, in order to produce a single isotherm. Simple weighted averaging is difficult to apply successfully because the different pores fill at different pressures, leading to a stepped isotherm. We circumvent this problem by fitting our simulated data to an empirical equation of state for a water/active carbon system. Fitting our directly obtained isotherms to this equation, and then interpolating the calculated empirical constants to all other pore-widths allows us to generate a single adsorption isotherm which is in reasonable quantitative accord with the experimental results. There are, however, several discrepancies, as discussed above. It is possible to attribute most of these inconsistencies to the interpolation equation used. However, improvement to the model of the surface sites is needed to give a more realistic picture of the low pressure region.

We have used a similar model to predict the adsorption behavior of ethane and ethylene on bare alumina and activated alumina. The model is able to quantitatively predict the adsorption behavior of these systems. The same molecular parameters have been used to predict the selectivity of ethylene versus ethane on the activated alumina at two different temperatures. As expected, the selectivity is enhanced when the temperature and pressure are lowered.

ACKNOWLEDGMENTS

We thank Ralph T. Yang for sending unpublished data. FJB thanks A. Z. Panagiotopoulos, A. D. Mackie and J. Bonet-Avalos for helpful discussions. Partial support was given by the Dirección General de Investigación Científica y Técnica (DGICyT, grant no. PB96-1025). International cooperation was made possible by a N.A.T.O. grant (no. CRG. 931517). We also thank the Department of Energy (grant no. DE-FG02-88ER13974) for support of this work and the National Science Foundation for a Metacenter grant (MCA935011P) which provided supercomputer time. One of us (SCM) thanks the NSF for a CISE Postdoctoral Fellowship. FJB thanks the Comissionat per a Universitats i Recerca de la Generalitat de Catalunya for a doctoral fellowship.

References

[1] K.S.W. Sing, D.H. Everett, R.A.W. Haul, L. Moscou, R.A. Pierotti, J. Rouquerol, and T. Siemineiewska, Pure Appl. Chem. **57**, 603 (1985).

[2] V.Y. Antonchenko, A.S. Davidof, and V.V. Ilyin, Physics of Water, (Naukova Dumka, Kiev, 1991).

[3] D.E. Ulberg and K.E. Gubbins, Molec. Sim. **13**, 205 (1994).

[4] D.E. Ulberg and K.E. Gubbins, Molec. Phys. **84**, 1139 (1995).

[5] M. Maddox, D.E. Ulberg and K.E. Gubbins, Fluid Phase Equil. **104**, 145 (1995).

[6] E.A. Müller, L.F. Rull, L.F. Vega, and K.E. Gubbins, J. Phys. Chem. **100**, 1189 (1996).

[7] P.B. Balbuena and K.E. Gubbins, Langmuir **9**, 1801 (1993).

[8] E.I. Segarra and E.D. Glandt, Chem. Eng. Sci. **49**, 2953 (1994).

[9] W.A. Steele, The Interaction of Gases with Solid Surfaces, (Pergamon, Oxford, 1974).

[10] R.T. Yang and E.S. Kikkinides, AIChE Journal **41**, 509 (1995).

[11] R.T. Yang and R. Foldes, Ind. Eng. Chem. Res. **35**, 1006 (1996).

[12] F.J. Blas, L.F. Vega, and K.E. Gubbins, Fluid Phase Equil. in press (1997).

[13] T.J. Bandosz, J. Jagiełło, and J.A. Schwarz, Anal. Chem. **64**, 891 (1992).

[14] J. Jagiełło, T.J. Bandosz, and J.A. Schwarz, Carbon **30**, 63 (1992).

[15] O. Talu and F. Meunier, AIChE Journal **42**(3), 809 (1996).

[16] R.C. Bansal and J. Donnet, in Carbon Black, edited by J. Donnet, R.C. Bansal, and M. Wang, (Marcel Dekker, New York, 1993).

[17] A.K. Soper and M.G. Phillips., Chem. Phys. **107**, 47 (1986).

[18] C.L. McCallum, T.J. Bandosz, S.C. McGrother, E.A. Müller, and K.E. Gubbins, in preparation (1997).

[19] W.L. Jorgensen, J.D. Madura, and C.J. Swenson, J. Am. Chem. Soc. **106**, 6638 (1984).

[20] L.E. Cascarini de la Torre, E.S. Flores, J.L. Llanos, and E.J. Bottani, Langmuir **11**, 4742 (1995).

[21] Y.-C. Xie and Y.-Q. Tand, Advances in Catalysis **37**, 1 (1990).

[22] N. Metropolis, A.W. Rosenbluth, M.N. Rosenbluth, A.H. Teller, and E. Teller, J. Chem. Phys. **21**, 1087 (1953).

SURFACE CHARACTERIZATION OF W/Ni/Al$_2$O$_3$ CATALYSTS

M.H. Jordão*, J.M. Assaf*, P.A.P. Nascente**

* Departamento de Engenharia Química, Universidade Federal de São Carlos, 13565-905 São Carlos, SP, Brazil, pmhj@iris.ufscar.br
** Centro de Caracterização e Desenvolvimento de Materiais, Departamento de Engenharia de Materiais, Universidade Federal de São Carlos, 13565-905 São Carlos, SP, Brazil, nascente@power.ufscar.br

ABSTRACT

Catalysts containing tungsten and nickel oxides are important in hydrodesulfurization (HDS), hydrogenation (HY), and steam reforming of hydrocarbons. A series of W/Ni/Al$_2$O$_3$ catalysts was prepared by two different methods: (1) coprecipitation of nickel and aluminium hydroxicarbonate from their nitrates, followed by calcination and impregnation of tungsten; (2) precipitation of boehmite from aluminium nitrate, followed by impregnations of nickel, firstly, and tungsten. The nickel content was kept constant, while the amount of tungsten varied from 2.5 to 15.5 wt-%. The resulting oxides were characterized by inductively coupled plasma spectroscopy (ICP), atomic absorption spectroscopy (AAS), X-ray diffraction (XRD), temperature programmed reduction (TPR), and X-ray photoelectron spectroscopy (XPS). ICP and AAS were used to determine the W, Ni, and Al concentrations. XRD detected two phases: NiO and NiAl$_2$O$_4$ (no phase containing metallic tungsten was detected). Increasing the amount of W, the quantity of NiAl$_2$O$_4$ rose, the quantity of NiO decreased, and the particle size of NiO enlarged. The TPR profiles presented three peaks: one at about 1000 °C, associated to a very stable phase; for the samples prepared by coprecipitation, the other two peaks corresponded to "free NiO" and a nonstoichiometric aluminate. For the samples prepared by impregnation, those peaks corresponded to NiO and NiAl$_2$O$_4$. XPS identified Al$_2$O$_3$, NiAl$_2$O$_4$, and Al$_2$(WO$_4$)$_3$ for both preparation methods. Increasing the amount of tungsten in the impregnated samples, NiWO$_4$ was also observed.

INTRODUCTION

Catalysts of nickel supported on alumina are applied to several industrial chemical reactions, such as hydrogenation and steam reforming of hydrocarbons. Catalysts of tungsten supported on alumina are also employed in hydrogenation reactions [1, 2], although nowadays their main application is in hydrodesulfurization [3, 4]. In this case, sulfated W-based catalysts are used [5, 6]. Catalysts of nickel supported on alumina with the addition of tungsten and alkaline metals as promoters have been studied in steam reforming of hydrocarbons, in which tungsten plays a fundamental role in the selectivity of some reactions [7-9].

In this work we have investigated the influence of tungsten on catalysts of nickel supported on alumina prepared by two methods: (1) coprecipitation of nickel and aluminium hydroxicarbonates from their nitrates, followed by calcination and impregnation of tungsten; (2) precipitation of aluminium hydroxide (boehmite) from its nitrate, followed by calcination and impregnations of nickel, firstly, and tungsten. The nickel content was kept constant, while the amount of tungsten varied from 2.5 to 15.5 wt-%. The W/Ni/ Al$_2$O$_3$ catalysts were characterized by several techniques: inductively coupled plasma spectroscopy (ICP), atomic absorption spectroscopy (AAS), X-ray diffraction (XRD), temperature programmed reduction (TPR), and X-ray photoelectron spectroscopy (XPS).

EXPERIMENTAL

We used two methods in the preparation of the W/Ni/ Al_2O_3 catalysts. The first method, designated here as COP, was the coprecipitation of nickel and aluminium hydroxicarbonates from their nitrates, followed by calcination to obtain Ni and Al oxides and impregnation of tungsten [10]. The second one, designated as IMP, involved the impregnation of nickel and tungsten on alumina. The coprecipitation was carried out at pH 8.0 and 60 °C, using sodium carbonate as precipitate agent [11]. The material was washed until sodium had been eliminated, dried, and calcinated in synthetic air at 500 °C. Then tungsten was impregnated, with excess of solvent, from silicotungstic acid. The nickel content was kept constant, while the amount of tungsten varied from 2.5 to 15.5 wt-%. The conditions and reactants used in the IMP process were analogous to those used in the COP process, differing only in the methodology. In the IMP case, nickel and tungsten were impregnated on alumina previously prepared by precipitation. The amount of tungsten also varied from 2.5 to 15.5 wt-%, while the nickel content was kept constant.

The quantities of Ni, Al and W in the resulting oxides were determined by inductively coupled plasma spectroscopy (ICP) and atomic absorption spectroscopy (AAS). The analyses by ICP were performed using an spectrometer Thermo Jarrel Ash (AtomScan 25), and those by AAS, by an spectrometer Intralab (AA 1475). The phase identification was done by X-ray diffraction (XRD) using a Carl Zeiss Jena (URD6) diffractometer. We employed the powder method using Cu Kα radiation (λ = 1.5405 Å), scanning velocity of 2°/min, and 2θ range of 5° to 90°. The particle size was calculated by the Scherrer equation. The determination of the atomic ratios and the identification of the oxidation states were carried out by X-ray photoelectron spectroscopy (XPS) using a Kratos (XSAM HS) spectrometer. The experiments were performed using Mg Kα radiation (hν = 1253.6 eV), under 10^{-9} Torr. The spectra were referenced to the binding energy of 284.8 eV for the C 1s line of adventitious carbon.

RESULTS AND DISCUSSION

The concentrations (weight percentages) obtained by ICP and AAS are presented in Table I.

Table I. Amounts of Ni, Al and W (in weight %) determined by ICP and AAS.

Sample	Ni	Al	W
COP Ni/Al	34.4	10.6	0
COP 2.5%	32.0	9.8	2.4
COP 6.5%	31.4	9.7	5.0
COP 9.5%	29.9	9.2	8.8
COP 15.5%	27.8	8.5	13.9
IMP Ni/Al	17.2	14.8	0
IMP 2.5%	16.5	14.2	2.5
IMP 6.5%	15.5	13.4	5.2
IMP 9.5%	14.7	12.7	9.6
IMP 15.5%	13.7	11.8	14.5

The diffractograms of the COP samples calcinated at 800 °C are presented in Figure 1. The catalyst precursor (# 1 in Figure 1) has hydrotalcite-type structure. The results indicate a phase mixture of NiO and $NiAl_2O_4$. Adding tungsten caused a shift in the (4 0 0) interplanar distance of the spinel structure, $NiAl_2O_4$, to interplanar distance values closer to the stoichiometric structure, indicating that this phase becomes more stable with the addition of tungsten. No metallic W was detected.

The particle sizes were determined by the deconvolution of the most intense XRD peak. For the COP samples, the particle size of $NiAl_2O_4$ remained approximately constant as the amount of W varied, while the particle size of NiO increased significantly (60-110 Å) as the amount of W rose.

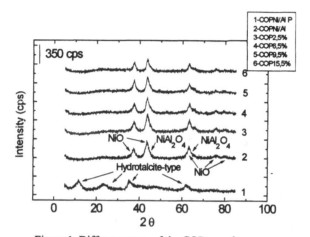

Figure 1. Diffractograms of the COP samples

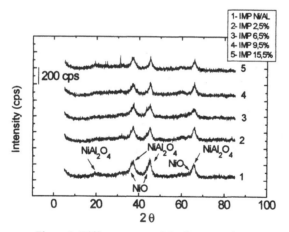

Figure 2. Diffractograms of the IMP samples

245

The impregnated catalysts were prepared using alumina support which had been previously made by the precipitation method. This alumina was identified as γ-Al_2O_3. Adding tungsten did not cause any change in the diffractograms (Figure 2). Again no metallic W was observed. The particle sizes of the IMP samples for both NiO and $NiAl_2O_4$ decreased with the increase of W amount.

The Figures 3 and 4 show that adding tungsten (2.5 w-%) increased the thermal stability of nickel compounds, with a shift of the main peak for coprecipited and impregnated catalysts up to 48 °C and 20 °C, respectively.

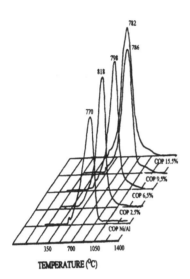

Figure 3. TPR of the COP samples

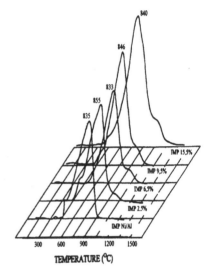

Figure 4. TPR of the IMP samples

The reduction temperature of the COP catalysts was increased as a consequence of the tungsten addition, although in the IMP catalysts this temperatures remains constant at 840 °C. Thus after the initial increase in the thermal stability, a competition occurs between Ni and W to form stable compounds with the aluminium. The increase in the W content results in easily reducible Ni phases, but with a higher thermal stability compared with non-tungsten catalysts.

A high temperature shoulder (900 °C and higher) is observed for the COP and IMP catalysts. The importance of this peak increases with the W loading and is best evidenced in the impregnated catalysts. An alumina-tungsten phase, such as $Al_2(WO_4)_3$, is probably formed, once this structure needs higher reduction temperatures than those for tungsten oxides [12].

The XPS results for the binding energies and atomic ratios are presented in Table II and Table III, respectively.

Table II. Binding energies (eV) for the COP and IMP catalysts.

Catalyst	Binding Energy (eV)		
	Al 2p	Ni $2p_{3/2}$	W $4f_{7/2}$
COP Ni/Al	75.0	856.5	—
COP 2.5%	75.4	857.1	36.9
COP 9.5%	75.2	856.8	36.6
IMP 2.5%	75.5	857.2	36.8
IMP 9.5%	74.1	855.9	35.4

Table III. Atomic ratios for the COP and IMP catalysts.

Catalyst	Ni/Al	W/Al	W/Ni
COP Ni/Al	0.43	—	—
COP 2.5%	0.56	0.029	0.052
COP 9.5%	0.50	0.088	0.17
IMP 2.5%	0.068	0.012	0.092
IMP 9.5%	0.25	0.095	0.38

The binding energy values of Ni $2p_{3/2}$ are consistent with $NiAl_2O_4$, but not with NiO; those of W $4f_{7/2}$ can be associated with $Al_2(WO_4)_3$ and WO_3 (in the case of IMP 9.5%). The values of Al 2p are in accordance with Al_2O_3, $Al_2(WO_4)_3$ and $NiAl_2O_4$. Comparing the values obtained for both COP 2.5% and IMP 2.5% catalysts, we observe that they correspond to the same compounds, indicating that the preparation method did not affect the magnitude of the nickel-support interaction.

The atomic ratios shown in Table III indicate that Ni and Al ions are distributed evenly, but the addition of a small quantity of tungsten causes a competition between Ni and W ions with Al ions, resulting in the surface segregation of Ni ions. The decrease in the Ni surface concentration as the amount of W increases indicates that a tungsten coating is forming on nickel for both series.

CONCLUSIONS

XRD identified only a mixture of NiO and $NiAl_2O_4$ phases for both COP and IMP catalysts. Using the deconvolution of the most intense XRD peak for determining the particle sizes, it was observed that the particle size of $NiAl_2O_4$ remained approximately constant with the increase of the W amount in the COP catalysts, while the particle size of NiO increased significantly. The particle sizes of the IMP samples for both NiO and $NiAl_2O_4$ decreased with the amount increase of W.

A shift in the interplanar distance of the spinel phase, $NiAl_2O_4$, to values closer to the stoichiometric structure was observed by XRD as tungsten was added to the COP catalyst, indicating a possible ordering of this metastable phase with the increase of the W amount. On the other hand, no similar shift was detected for the IMP samples.

The XPS results indicate the presence of $NiAl_2O_4$ and $Al_2(WO_4)_3$ in the surface of both COP and IMP catalysts with different amounts of W, and WO_3 (in the case of IMP 9.5%). Binding energies of Al, Ni and W for both COP 2.5% and IMP 2.5% catalysts correspond to the same compounds, indicating that the preparation method did not affect the degree of the nickel-support interaction.

ACKNOWLEDGMENTS

This work was supported by the Brazilian agencies FAPESP and CNPq.

REFERENCES

1. R. Thomas, E. M. Van Oers, V. H. J. de Beer, J. Medema and J. A. Moulijn, .Journal of Catalysis **76**, p.241 (1982).
2. E. Kural, N. W. Cant, D. L. Trimm and C. Mauchausse, J. Chem. Tech. Biotechnol. **50**, p. 493 (1991).
3. W. Grünert, E.S Shpiro, R. Feldhaus, K. Anders, G. V. Antoshin and Kh. M. Minachev, Journal of Catalysis **107**, p.522 (1987).
4. P. J. Mangnus, A. Bos and J. A. Moulijn, Journal of Catalysis **146**, p.437 (1994).
5. B. Scheffer, J. J. Heijeinga and J. A. Moulijn, J. Phys. Chem. **91**, p. 4,752 (1987).
6. P. Arnoldy, M.C. Franken, B. Scheffer and J.A. Moulijn,. Journal of Catalysis **96**, p.381 (1985).
7. L. Bonneau, K. Arnaout, and D. Duprez, Applied Catalysis. **74**, p. 173 (1991).
8. G. Delahay, J. Bousquet and D. Duprez, Bull. Soc. Chim. F. II **6**, p.1,237 (1985).
9. G. Delahay, and D. Duprez, Bull. Soc. Chim. F. II **6**, p.1,245 (1985).
10. J. Zielinski, Applied Catalysis A: General **94**, p. 107 (1993).
11 S. Narayanan, R. Unnikrishnan and V. Vishwanathan, Applied Catalysis A: General **129**, p.9 (1995).
12. A. A. Van Roosmalen and J. C. Mol, Journal of Catalysis **78**, p.17 (1982).

AUTHOR INDEX

SUBJECT INDEX

Printed in the United States
By Bookmasters